Levelling Up the

"This excellent book provides a powerful, cr[illegible] examination of Levelling Up, and grounds its assessment within a much-needed political economy perspective. It traces the 'left behind places' problem to the nature and evolution of UK capitalism itself. Further, it goes beyond the usual aggregate statistical metrics used to measure and discuss the scale of the 'left behind places' problem, to delve into the lived experiences of those living in those places. By means of this political economy and qualitative approach, the authors rightly conclude that the UK Government's Levelling Up programme is unlikely to deliver on its promises to transform economic fortunes and social conditions in 'left behind places', and may not even prevent geographical inequalities from widening still further. This stimulating book is essential reading for all academic scholars and policymakers concerned with the UK's 'left behind places problem'."

—Professor Ron Martin, *Emeritus Professor of Economic Geography, University of Cambridge, UK*

"This is one of the most significant efforts to analyse the UK government's Levelling Up programme. Not only does it brilliantly chronicle the moral, social and economic reasons for addressing geographical inequality, but it further details the transformative changes necessary to overcome the significant barriers that people in 'left behind' locales face in terms of gaining access to well-paid, secure employment. Recognising that spatial disparities in the UK are deeply entrenched and long-running, Telford and Wistow argue for a 'phase shift' in political economy that goes beyond the 'sticking plaster' approach to the present, which they suggest is unlikely to spread opportunity, revive communities and restore local pride. This book is likely to garner much interest from those interested in trajectories of inequality and the implications for place, people and government policy."

—Professor Julie MacLeavy, *Professor of Economic Geography, University of Bristol, UK*

Luke Telford • Jonathan Wistow

Levelling Up the UK Economy

The Need for Transformative Change

palgrave
macmillan

Luke Telford
Department of Social Work and Social Policy and School for Business & Society
University of York
York, UK

Jonathan Wistow
Department of Sociology and Wolfson Research Institute for Health and Wellbeing
Durham University
Durham, UK

https://doi.org/10.1007/978-3-031-17507-7

This Palgrave Macmillan imprint is published by the registered company Springer Nature Switzerland AG.
The registered company address is: Gewerbestrasse 11, 6330 Cham, Switzerland

Foreword

I am writing this on the day that the Conservative Government has announced new 'investment zones' across the UK. These are to be based on slashing personal taxes in addition to business rates as part of a national wheels in motion economic strategy, which builds on previous policy experiments such as enterprise zones, development corporations, and free-ports. Previous obsessions with subnational economic development as part of what was called the 'Levelling Up' agenda appear to be being superseded by a national growth agenda. This is looking more like Hayek on helium—an ultra-faith in free markets, laissez-faire, and a belief that business knows best, and wealth will indeed trickle-down to our cities and regions.

Academics and policymakers alike need to read Levelling Up the UK Economy by Luke Telford and Jonathan Wistow. This book is the most comprehensive qualitative study available on the 'levelling up' chapter in the policy history of local and regional economic development, which occupied the period 2019–2022. Here, the authors skillfully work through 30 semi-structured interviews with 5 Directors of Regeneration working in so-called left behind places, as well as unique insight offered by the lives of 25 'left behind' residents in Teesside—an important experimental space for the Conservative Government's ambitions for free-market capitalism. Theoretically informed, interdisciplinary, and rich in policy nuances and political insight, the authors walk us through the rise and fall of this policy narrative and argue critically that rather than providing a major break in the patterns of combined and uneven development in the UK's polity, Levelling Up should be seen instead as tinkering at the edges of transformative change. Yes, there has been a renewed focus on pride in policy and civic collective visions, but the event to which this has really empowered local government and the fabric of civil society is open to critical question. This is social policy at its very best.

Rather than just leave things there, the authors conclude by offering an ambitious alternative path for our 'left behind places'. Now that 'Levelling Up' is predicted to be eclipsed by a new national growth agenda, the authors persuasively invite readers to consider the possibilities for a Green Industrial Revolution, where deglobalization is a core component of the UK political economy. This is pitched alongside interventions such as a state funded job guarantee, argued to be pivotal in transforming those 'left behind' places such as Teesside. Writing this foreword from my university base in Stoke-on-Trent—not too dissimilar in economic and social challenge to Teesside—I welcome these initiatives which need to be subjected to serious academic and policy debate.

Vice Chancellor and Professor of Human Geography, Staffordshire University
Stoke-on-Trent, UK

Martin Jones

Acknowledgements

The authors' ongoing collaborative work stems from Jonathan supervising Luke's MA project at Durham University back in 2017. This project brings together much of our thinking in the years since, synthesising the government's flagship policy programme—Levelling Up—particularly with insights from politics, political economy, sociology and economic geography. There are too many people to thank, though Luke is particularly thankful to his family especially his Mum as well as his wife Gemma, who he dedicates the book to. Thanks also go to his Labrador, Freddie, for time away from the book on walks. Jonathan would like to thank Carol for all her support and patience and Benjamin for being a good lad! We also thank the respondents for giving up their time to speak to us, since without them the book's empirical contribution would not have been possible.

Contents

About the Authors

Luke Telford is a Lecturer in Criminal Justice and Social Policy at the University of York. Luke is an interdisciplinary researcher. His main interests include working-class culture, social harm, politics, political economy, the COVID-19 pandemic and developing social policy for 'left behind' places. He is the author/co-author of four books. This includes *Lockdown: Social Harm in the COVID-19 Era* (Palgrave Macmillan, 2021), which explored the unintentional consequences of pandemic management such as the lockdowns and associated restrictions. It also includes the monograph *English Nationalism and Its Ghost Towns* (Routledge, 2022), exploring the factors that gave rise to political discontent in a 'left behind' place and what can be done to ameliorate it. Luke is working on the problems of 'left behind' places and what can be done to resurrect them in relation to social policy and the political economy.

Jonathan Wistow is an Associate Professor at Durham University. His core interests include place-based social policy and governance systems, political economy and health inequalities. He is the author/co-author of three books including *Studying Health Inequalities: An Applied Approach* (Policy Press, 2015), as well as a recent book entitled *Social Policy, Political Economy and the Social Contract* (Policy Press, 2022). The latter explores complexity and social contract theory to understand the trajectory of political economy and its linkages with policy. Jonathan is researching place-based health inequalities, air pollution and brain health, and Levelling Up 'left behind' locales all within a complex systems frame of reference.

List of Tables

CHAPTER 1

Introduction

Abstract This chapter provides a brief summary of the Levelling Up 'left behind' places problem. It also outlines some notes on the book's methodological underpinning including its complex systems frame of reference and how this shaped the qualitative data collection, encompassing 25 interviews with residents of 'left behind' Redcar & Cleveland in Teesside and 5 interviews with Directors of Regeneration from various 'left behind' localities. The chapter closes by outlining the structure of the book.

Keywords Levelling Up • 'Left behind' • Complex systems

The Levelling Up agenda was introduced during Boris Johnson's first speech as Prime Minister (PM's Office, 2019). It featured prominently in the 2019 general election campaign and forms a central pillar of the policy rhetoric and programme for the Conservative Government that won the election. Alongside the promise to 'get Brexit done', the notion of 'Levelling Up' apparently appealed to the traditional Labour Party voting constituencies of the so-called 'Red Wall' of 'left behind' places that had also generally voted to leave the EU in the 2016 referendum. Accounts of these political shifts highlight the significance of political abandonment, deindustrialisation, relative deprivation, welfare retrenchment and general anxieties about conditions and opportunities in the present and the future

L. Telford, J. Wistow, *Levelling Up the UK Economy*,
https://doi.org/10.1007/978-3-031-17507-7_1

(Eatwell & Goodwin, 2018; Tilley & Evans, 2017; MacLeavy & Jones, 2021; Telford & Wistow, 2020; Telford, 2022; Winlow et al., 2017). Whilst Brexit took effect on 31st January 2021 (albeit with ongoing issues with both the terms on which the UK left the EU and the implementation of these) it is less clear how Levelling Up will be achieved or, perhaps more importantly, to identify when it has been achieved. Both prior to and following the launch of the 2022 Levelling Up White Paper (LUWP), the agenda has attracted much attention as a flagship policy programme of the Johnson Government but with sustained criticism about its scale, focus and ambiguity (see, for example: Carr-West & Sillett, 2021; Connolly et al., 2021; House of Commons Business, Energy and Industrial Strategy Committee, 2021; Liddle et al., 2022). In this introductory chapter, we provide a very brief summary about Levelling Up, which is followed by a methodological note on how the policy programme is contextualised within a wider understanding of complex systems. The chapter concludes with an outline of the book structure.

The Levelling Up Agenda

The LUWP (HM Government, 2022, p. xii) states that Levelling Up means ensuring 'everyone shares equally in the UK's success', 'giving everyone the opportunity to flourish', which 'requires us to end the geographical inequality which is such a striking feature of the UK'. In the foreword to the White Paper (HM Government, 2022, p. viii), the then Prime Minister Boris Johnson claims the UK economy is so unbalanced that:

> a country in which the place of your birth is one of the clearest determining factors in how you'll get on, what opportunities will be open to you, even the number of years for which you can expect to live…the challenges we face have been embedded over generations and cannot be dug out overnight.

This view is broadly consistent with the literature and evidence about spatial disparities in the UK (for example: Jones, 2019; MacLeavy & Manley, 2019; Martin, 1988; Martin et al., 2021; Pike, 2022; Wistow, 2022), posing some fundamental challenges to the nature and direction of the UK's political and spatial economy. In attempting to level up places the government (with a good deal of ambiguity built in) is essentially seeking to create a more equal playing field in terms of spatial conditions, opportunities, outputs and outcomes. However, this occurs in a context in which the cumulative effects of managing the transition to a post-industrial economy (Beatty & Fothergill, 2016; Byrne, 2019; Martin, 2015) within a broadly neoliberal (Crouch, 2011); rentier (Christophers, 2020; Lansley, 2022); post-democratic (Crouch, 2020); and global (Castellani, 2018) political economy has created a trajectory and path dependency that has intersected across place and the socio-economic realities of individuals' lives. The manifestation of spatial inequalities across this trajectory have taken their toll on communities and are a key and sustained causal factor relative to social outcomes and opportunities in place, including for example the labour market, housing market, educational quality and outcomes, health inequalities, through to the experience of and outcomes resulting from the COVID-19 pandemic and associated restrictions. As we will see, there are serious doubts that the Levelling Up agenda will be either equal to the challenge and policy rhetoric or sufficiently radical to fundamentally shift the trajectory and path dependency of spatial disparities in the UK.

The key problem that the Levelling Up agenda seeks to address is that of 'left behind places.' Martin et al. (2021, p. 7) describe 'the "left behind" problem as being 'spatially and systematically entrenched', albeit with distinct trajectories at different scales, 'due to the inability of some places to adapt to the post-industrial service and knowledge-based economy'. In effect, the Levelling Up agenda is a policy programme that is intervening in the trajectory of social, political and economic systems to shape the spatial configuration of largely economic and social outcomes. This, in turn, poses the challenge of understanding systems, their complexity and how they intersect with one another (Castellani, 2018). In so doing, we need to acknowledge context and the mutual influence of macroscopic and microscopic structures and dynamics in social systems (Byrne & Callaghan, 2014). In this respect, Wistow (2022, p. 51) asserts that 'the political economy has a powerful position within the nexus (the system of systems) that in complexity theory is viewed as constituting the make-up and interactions between social, political, economic, cultural and institutional domains cutting across complex systems.' He (2022) continues to argue that the influence of (an albeit highly selective approach to) neoliberalism has considerable impact over the nature and extent of social outcomes (of the type described in the Levelling Up agenda), while also influencing the capacity and type of policies designed to respond to these.

In other words, the 'left behind' problem is a product of a highly unequal and centralised neoliberal and rentier political economy, intersecting across systems at different spatial levels. The way the economy is thus managed including interventions through various levels of state, market, and civic engagement should be fully embedded into policy responses seeking to address spatial disparities. As Jones (2019, p. 109) argues, 'in the absence of significant national state intervention, capitalism tends towards spatially uneven development,' while Martin et al. (2021, pp. 76–77) similarly concludes 'without intervention, places with depressed economies have been found to suffer an erosion of the capacities needed to sustain resilience. A good example is the recession of the 1980s which was felt most strongly in Britain's manufacturing heartlands.' They (2021) continue to highlight that the decision-making of national and local economic and political actors accentuates or ameliorates the impacts of economic shocks. Broadly speaking, the tendency in the UK has been to accentuate the spatial imbalances of economic shocks through a political economy that favours state intervention in the interests of financial capital rather than the social and economic interests of the public.

Methodological Note

Before we outline the structure of this book it is necessary to provide a brief methodological note on the framing of the content and how different strands of primary and secondary data will be employed. Our ontological and epistemological backgrounds primarily lie in complexity theory, critical realism and ultra-realism.[1] These are complementary positions that align well with the focus on place in the Levelling Up agenda and the social and economic reality of the spatial disparities it seeks to address. Place is highly significant for a range of social and economic outcomes (see, for example: MacLeavy & Jones, 2021; Martin, 2015; HM Government, 2022; UK2070 Commission, 2020; Telford, 2022; Warren, 2018), constituting a key component of the material reality that shapes our experience and quality of life. Place also draws attention to a range of dynamic interactions across spatial hierarchies (e.g., at the local level within the global and national political economy); cross-sectoral policy

[1] Theoretically Luke is an ultra-realist, which is a paradigm that has been somewhat influenced by critical realism (see: Bushell, 2022; Hall & Winlow, 2015; Kotze & Lloyd, 2022; Raymen & Kuldova, 2021). Luke has utilised ultra-realist theoretical ideas in much of his work in recent years (for example: Briggs et al., 2021; Telford & Lloyd, 2020; Telford, 2022; Telford et al., 2022).

regimes and governance systems; and temporal dimensions (e.g., post-industrial transitions). In response the LUWP (HM Government, 2022) explicitly contextualises the agenda in chapter 2 around geographical disparities over time and across different spatial scales, while in chapter 3 it frames its response in terms of systems reform. How appropriately and effectively the agenda captures and responds to these issues will be a key concern of the book. For now, we want to develop our approach to understanding the Levelling Up agenda as a policy programme within a complex systems framing of policy and the political economy.

In seeking to understand the complexity described above we use different 'lenses'[2] (Head, 2008) and 'entry points'[3] (Jessop, 2016) to explore the Levelling Up agenda as: a policy discourse and strategy at national and local levels; as it relates to a range of regional and local social and economic outcomes; and as it relates to the lived experiences of residents in a so-called 'left behind' place—Redcar & Cleveland (R&C) in Teesside. Our approach employs contextual data about place at different spatial levels of aggregation, academic and grey literature, official reports, and expert witness oral and written evidence to, for example, parliamentary select committees. For instance, Chap. 2 develops a historical framing of path dependencies and trajectories over time, often focussing on R&C and Teesside as both a case for analysis and for how locality links to other systems. This local authority is also the focus of Chap. 5, which draws on original primary data from interviews with working class residents. In Chaps. 3 and 4 we also draw on primary data from scoping discussions with local authority Directors of Regeneration in so-called 'left behind' areas, which provides a further lens through which the study was conducted. We used Martin et al.'s (2021) classification of 'left behind' local authorities in the bottom quartile for cumulative growth in employment and outputs between 1981 and 2018 to identify participating areas for the study. In so doing the trajectory and path dependency of place across the neoliberal and post-industrial UK political economy is factored into the

[2] Head (2008) focuses on the inherently political and value-based nature of policy debate and decision-making that, he argues, leads to policy decisions being 'deduced not primarily from facts and empirical models, but from politics, judgement and debate.' Head (2008) suggests that there are three lenses of evidence-based policy: political knowledge; scientific (research-based) knowledge; and practical implementation knowledge.

[3] Following Jessop (2016) policy areas and policies can be employed as specific 'entry points' to identify and understand divergent perspectives about the role of the state and associated organisations and actors.

complexity frame of reference, given the longitudinal dimension of this classification.

The primary data collection included a small scoping study involving semi-structured interviews between May and July 2022 with five Local Authority Directors of Regeneration featuring in Martin et al.'s (2021) 'left behind' areas. They are classified as: a 'Northern Core City'; a 'Northern Other City'; a 'Northern Large Town'; a 'Midlands Medium Town'; and a 'London Borough'. This scoping study was accompanied with 25 semi-structured interviews with working class individuals in post-industrial R&C in Teesside, which has been identified by Martin et al. (2021) as the 2nd most 'left behind' local authority areas out of 74 in terms of the differential growth of both employment and output. These interviews were conducted between May and August 2022 with 22 conducted in private residence, pubs and cafes and the remaining three over the telephone. They generally lasted around 35 minutes. Respondents ranged from age 21–80, with the majority aged above 40 including nine who were retired. They were able to offer more long-term views on the area's socio-economic decline; how it has changed in recent decades and the implications for the Levelling Up strategy. 11 respondents are female, the rest male. All respondents are white, which reflects local demographics as R&C contains the highest proportion of White British residents in the UK. Most would identify as working class, though some lived in R&C's more affluent neighbourhoods. Therefore, our small-scale qualitative study builds upon the dearth of qualitative research in 'left behind' locales, responding to MacKinnon et al.'s (2022, p. 40) suggestion that 'research has yet to fully engage with the development problems of such 'left behind' places' particularly in relation to Levelling Up. Our aims are modest—to empirically contribute to the growing social, academic and political debates on 'left behind' places and Levelling Up.

In order to contextualise the Levelling Up agenda, we have developed Fig. 1.1 which provides a 'non-exhaustive' complex systems map that acts as a multi-scalar overview of the policy agenda, key actors and systems logic. This is not a complete account, though it provides an insight into how a range of processes, trends, dynamics and experiences interact across spatial contexts and scales. The systems map has been developed through drawing on a wide range of literature and has been modified using insights from the qualitative data collected and analysed in Chaps. 3, 4 and 5. In particular, literature from: economic geography (e.g., Beatty & Fothergill, 2016; MacLeavy & Manley, 2019; Martin et al., 2021); post-industrialism

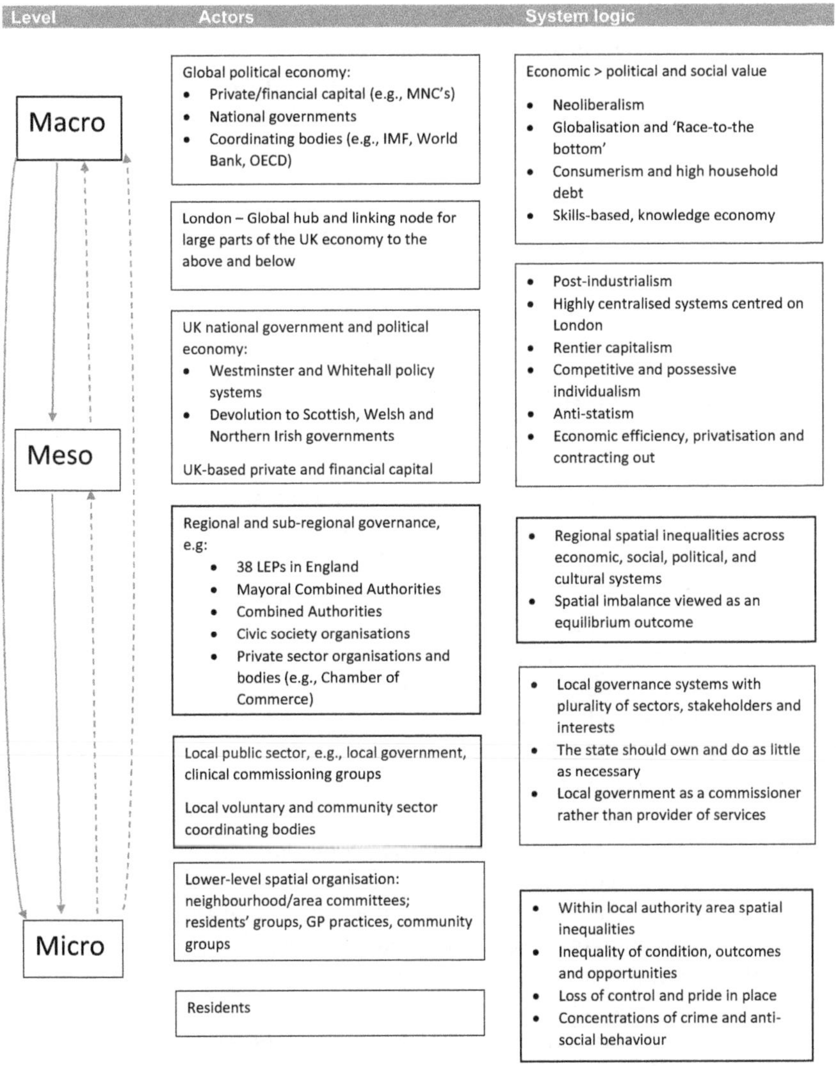

Fig. 1.1 A 'non-exhaustive' Levelling Up complex systems map

and class (e.g., Byrne, 2019; Telford, 2022; Warren, 2018); political economy (Christophers, 2020; Streeck, 2016) complexity theory (Castellani et al., 2015; Haynes, 2003); governance and policy systems (Byrne &

Callaghan, 2014; Wistow, 2022); and local and regional government studies (e.g., Barnett et al., 2021; Jones, 2019) has been used in the development of Fig. 1.1. The complex systems map provides a connecting framework for the book across three areas that are discussed briefly below.

Level

The first column in Fig. 1.1 relates to a multi-level understanding of complex systems. Byrne's (2005, p. 105) conceptualisation of cases in the complexity frame is useful here, encouraging us to view cases as 'complex systems which are nested in, have nested within them, and intersect with other complex systems.' As Byrne (2005) continues to argue, these levels have implications for each other through the ways they intersect and constitute one another, which is illustrated across the two remaining columns in the figure. Given the focus on place and significance of political and economic processes at different levels for regional and local spatial disparities, this type of approach to complexity is highly relevant to the Levelling Up agenda. For example, Castellani (2018) argues that the global capitalist system takes the form of a worldwide negotiated ordering of competing individuals and groups seeking happiness, often at the expense of others, with those at the wrong end of inequality or economic ruin tending to be deeply affected by the global commitments others have. As we will see, the macro has considerable implications for the system logic cutting across complex systems at lower levels of aggregation. There is agency at all levels outlined in Fig. 1.1, but we follow Verkerk et al. (2015) in viewing this as a hierarchical system in which actors interact in subsystems at multiple levels. As Srnicek and Williams (2016) argue, large-scale systemic problems occur in local communities but reducing them to this level of analysis denies the interconnections of the social world. We must also add the roll of positive and negative feedback between levels, with the latter being particularly significant relative to the 'left behind' problem and the aim to Level Up. Haynes (2003) describes negative feedback loops requiring maximum effort and action to break the cycle and start a positive trajectory. The Levelling Up agenda is an attempt at doing just that.

Actors

Figure 1.1 also includes details of key actors at different spatial levels. For example, the macro includes multinational corporations (MNCs), the meso Local Enterprise Partnerships (LEPs) and the micro local residents. A key feature of the British system is policy reinvention (Jones, 2019), which has contributed to the changing landscape of local governance and the actors and interdependencies comprising them. Martin et al. (2021, p. 100) describes this as a 'proliferation and churn' of spatial policy schemes and bodies with 52 being introduced between 1978 and 2018. This churn occurs within, and plays a significant role in constituting, multi-level governance (MLG). Broadhurst and Gray (2022, p. 5) argue 'that multi-level governance describes the dispersion of power from central governments to other 'centres' and the sharing of policymaking responsibility between supranational, national, regional and local government. As such, MLG theory seeks to understand the complex array of actors and processes involved in policy making.' A second key feature of the system is that the UK has, in Skelcher's (2000) terms, become a 'congested state' through the onset of governance. In Chap. 4 we focus in particular on the role of local government within local governance systems. Jones (2019, p. 217) argues that this level of analysis 'has to be set within the context of the mixture of local institutions, their scalar/territorial relations, and the associated dynamics of class and political struggle.' Our approach, as illustrated through Fig. 1.1, does just that using the micro as our main 'entry point' and route into explaining large-scale systemic problems relative to the role of government in Chap. 4, which is extended to the neighbourhood level of the micro in Chap. 5 through residents' views and experiences.

System Logic

The third column of Fig. 1.1 identifies some key elements of the system logic at different levels of spatial aggregation. Given the focus on complexity it is important to stress that these are not intended to be deterministic. Nevertheless, there are path dependencies within the trajectory of the political economy that should be considered within an ideological lens, which has significance when applied as a system logic at different scales. As Castellani et al. (2015, p. 51) argue, 'the dynamics of most places, while stochastic, do evolve along a reasonably stable set of trajectories (attractor points).' The system logic developed here, therefore, is hierarchical, involving agency at different levels. At the macro level of Fig. 1.1 the global and national political economy has tended towards neoliberalism, rentier capitalism, post-industrialism, and post-democracy. We will deal with these more fully throughout the book. For now, Lansley (2022, p. 116) summarises this as 'involving dismantling parts of the policy apparatus constructed to ensure equal societies and minimise poverty and preparing the conditions for the forward march of the super-rich. Keynes and Titmuss were out. Hayek and Friedman were in.' He continues to argue that the post-war extension of social, economic and civic rights through an enhanced safety net, expansion of public goods and services, and access to decent training, work and wages, alongside business elites accepting wider responsibilities have been eroded throughout neoliberalism. Wistow (2022) describes this as a 'phase shift' in the political economy, the complex governance and policy systems influenced by this, and the competitive individualism and inequalities it has amplified.

In short, the macro-trajectory has a path dependency that has influenced the meso and micro levels in this period. The meso and micro have significant feedback, dynamics and agency but as argued above the overarching system logic is hierarchical, not least through the highly centralised nature of the UK political, economic, policy and cultural systems, which all have key nodes concentrated in London. In turn, Jones (2019, pp. 30–31) describes how the 'inherent contradictions and crisis tendencies within capitalism are governed within economic development and through a historically variable set of institutional, spatial-temporal, and semantic fixes—all partial, provisional and temporary.' Haynes' (2003) identification of 'entanglement' and dependence between open systems is useful here, since spatial inequalities in the UK can be viewed as a systemic feature of a highly centralised, neoliberal and rentier political economy. At the meso level, spatial disparities have intensified from the 1980s onwards (MacLeavy & Manley, 2019; Martin et al., 2021; Pike, 2022), meaning we should consider this as a form of entrenched negative feedback for many localities across the nexus. Martin et al. (2021) summarises this as a major phase of geographically divergent development and historic systemic transformation, which provides the political economic backdrop to 'left behind' places and Levelling Up. They continue to situate the 'left behind' problem within the economic and cultural transition to post-industrialism, which has not been economically, socially or spatially neutral with places at different scales unable to adapt equally successfully to the changes involved. In short, a systems approach is essential to understand what is being 'Levelled Up' to and the potential to do this across entrenched and divergent spatial inequalities.

Book Structure

Chapter 2—capitalism's trajectories and local spatial dynamics—outlines the macro context of the UK political economy and its implications for place and spatial inequalities. It documents key political economic shifts from the nineteenth century into capitalism's post-war phase and then explores the shift to a post-industrial neoliberal political economy from the late 1970s onwards. The chapter also explores the implications of the 2008 global financial crash and the imposition of spatially regressive austerity measures. Throughout, it mentions the implications of these changes for R&C in Teesside, providing important context for the 'left behind' place case study explored in Chap. 5. The next chapter—the Levelling Up agenda—is organised into two parts. The first considers the development of the Levelling Up agenda prior to the launch of the 2022 White paper. The second section outlines key features of the White Paper and focuses in particular on 'systems reform', 'capitals' and 'missions'. Chapter 4—Local Government, Governance and Levelling Up—explores the Levelling Up agenda from the 'entry point' of local government, drawing on academic literature and a small number of scoping interviews from Directors of Regeneration in 'left behind' places. The chapter considers the role of local government in local governance systems, local democracy and devolution and concludes by returning to the implications of the Levelling Up agenda for the role of local government in tackling spatial disparities. The chapter also reintroduces some earlier themes such as deindustrialisation and austerity to explore them in relation to local governance.

Chapter 5—Sentiments from a 'left behind' place—offers qualitative data from working class residents in R&C. It begins by briefly exploring sentiments around the economic stability and security of industrial work and the implications of the shift to more service-based forms of employment under neoliberalism. Next, it addresses the impact of this contextual change upon community life including the decline of many local town centres, the prevalence of crime, and socially corrosive forms of anti-social behaviour (ASB). The chapter also explores debates around the 'North South' divide from the perspective of people living in R&C in Teesside. It closes with respondents' sentiments on Levelling Up, exploring what the agenda means to them and how they believe it can be a success. Chapter 6 is the concluding chapter of the book. It starts by summarising some of its key features, suggesting as it stands the Levelling Up agenda is a 'sticking plaster' given the deeply embedded and long-running spatial inequalities in the UK. It then explores some contemporary debates about changes to the political economy given the waning legitimacy of neoliberalism. Alternative strategies that would create a more conducive context to Level Up and represent a re-calibration of the political economy are then discussed. Throughout the book, the complex systems framework/language is often explicitly deployed, though at times particularly in Chap. 2 we simply document how the macro context of political economy shapes the meso and micro levels.

References

Barnett, N., Giovannini, A., & Griggs, S. (2021). *Local government in England: 40 years of decline*. Unlock Democracy.

Beatty, C., & Fothergill, S. (2016). *The uneven impact of welfare reform: The financial losses to people and places*. Sheffield: Centre for Regional Economic and Social Research.

Briggs, D., Telford, L., Lloyd, A., Ellis, A., & Kotze, J. (2021). Lockdown: Social harm in the Covid-19 era. London: Palgrave Macmillan.

Broadhurst, K., & Gray, N. (2022). Understanding resilient places: Multi-level governance in times of crisis. *Local Economy, 37*(1–2), 84–103.

Bushell, M. (2022). No time for rest: An exploration of sleep and social harm in the North East Night-Time Economy (NTE). *Critical Criminology*. Online First. https://doi.org/10.1007/s10612-022-09655-8

Byrne, D. (2005). *Social exclusion*. Open University Press.

Byrne, D. (2019). *Class after industry: A complex realist approach*. Palgrave Macmillan.

Byrne, D., & Callaghan, G. (2014). *Complexity theory and the social sciences: The state of the art*. Routledge.

Carr-West, J., & Sillett, J. (2021). *On the level: Six principles to underpin the levelling up White Paper*. London: LGiU.

Castellani, B. (2018). *The defiance of global commitment: A complex social psychology*. Routledge.

Castellani, B., Rajaram, R., Buckwalter, J., Ball, M., & Hafferty, F. (2015). *Place and health as complex systems: A case study and empirical test*. Springer.

Christophers, B. (2020). *Rentier capitalism: Who owns the economy, and who pays for it?* Verso.

Connolly, J., Pyper, R., & van der Zwet, A. (2021). Governing 'levelling up' in the UK: Challenges and prospects. *Contemporary Social Science, 16*(5), 523–537.

Crouch, C. (2011). *The strange non-death of neoliberalism*. Polity Press.

Crouch, C. (2020). *Post-democracy: After the crisis*. Polity Press.

Eatwell, R., & Goodwin, M. (2018). *National Populism: The Revolt against Liberal Democracy*. London: Pelican Books.
Hall, S., & Winlow, S. (2015). *Revitalizing criminological theory: Towards a new ultra-realism*. Routledge.
Haynes, P. (2003). *Managing complexity in the public services*. Open University Press.
Head, B. (2008). Three lenses of evidence-based policy. *Australian Journal of Public Administration, 67*(1), 1–11.
HM Government. (2022). *Levelling Up: Levelling Up the United Kingdom*. London: Her Majesty's Stationery Office.
House of Commons Business, Energy and Industrial Strategy Committee. (2021). *Third Report—Post-pandemic economic growth: Levelling up*. London: House of Commons.
Jessop, B. (2016). *The state: Past, present, future*. Polity Press.
Jones, M. (2019). *Cities and regions in crisis: The political economy of sub-national economic development*. Edward Elgar Publishing.
Kotze, J., & Lloyd, A. (2022). *Making sense of ultra-realism*. Emerald.
Lansley, S. (2022). *The richer the poorer: How Britain enriched the few and failed the poor. A 200-year history*. Policy Press.
Liddle, J., Shutt, J., & Addidle, G. (2022). Editorial: Levelling Up the United Kingdom? A useful mantra but too little substance or delivery? *Local Economy, 37*(1–2), 3–12.
MacKinnon, D., Kempton, L., O'Brien, P., Ormerod, E., Pike, A., & Tomaney, J. (2022). Reframing urban and regional 'development' for 'left behind' places. *Cambridge Journal of Regions, Economy and Society, 15*, 39–56.
MacLeavy, J., & Jones, M. (2021). Brexit as Britain in decline and its crises. *The Political Quarterly, 92*(3), 444–452.
MacLeavy, J., & Manley, D. (2019). Socio-political fracturing: Inequality, stalled social mobility and electoral outcomes. *Area, 51*, 681–688.
Martin, R. (1988). The political economy of Britain's North-South divide. *The Royal Geographical Society, 13*(4), 389–418.
Martin, R. (2015). Rebalancing the spatial economy: The challenge for regional theory. *Territory, Politics, Governance, 3*(3), 235–272.
Martin, R., Gardiner, B., Pike, A., Sunley, P., & Tyler, P. (2021). *Levelling Up left behind places: The scale and nature of the economic and policy challenge*. Routledge.
Pike, A. (2022). Coping with deindustrialization in the global North and South. *International Journal of Urban Sciences, 26*(1), 1–22.
Prime Ministers Office. (2019). Boris Johnson's first speech as Prime Minister. 24 July 2019.
Raymen, T., & Kuldova, T. (2021). Clarifying ultra-realism: A response to Wood et al. *Continental Thought & Theory, 3*(2), 242–263.

Skelcher, C. (2000). Changing images of the state: Overloaded, hollowed-out, congested. *Public Policy and Administration, 15*(3), 3–19.

Srnicek, N., & Williams, A. (2016). *Inventing the future: Postcapitalism and a world without work.* Verso.

Streeck, W. G. (2016). *How will capitalism end?* Verso.

Telford, L. (2022). *English nationalism and its ghost towns.* Routledge.

Telford, L., Bushell, M., & Hodgkinson, O. (2022). Passport to neoliberal normality? A critical exploration of COVID-19 vaccine passports. *Journal of Contemporary Crime, Harm, Ethics, 2*(1), 42–61.

Telford, L., & Lloyd, A. (2020). From 'infant Hercules' to 'Ghost Town': Industrial collapse and social harm in Teesside. *Critical Criminology, 28*, 595–611.

Telford, L., & Wistow, J. (2020). Brexit and the working class on Teesside: Moving beyond reductionism. *Capital & Class, 44*(4), 553–572.

Tilley, J., & Evans, G. (2017). The New Politics of Class after the 2017 General Election. *The Political Quarterly, 88*, 710–715. https://doi.org/10.1111/1467-923X.12434

UK2070 Commission. (2020). *Go big. Go local: The UK2070 report on a new deal for levelling up the United Kingdom.* Nottingham: UK2070.

Verkerk, J., Teisman, G., & Van Buuren, A. (2015). Syncronising climate adaptation processes in a multilevel governance setting: Exploring synchronisation of governance levels in the Dutch Delta. *Public Management Review, 43*(4), 579–596.

Warren, J. (2018). *Industrial Teesside, lives and legacies.* Palgrave Macmillan.

Winlow, S., Hall, S., & Treadwell, J. (2017). *The Rise of the right: English nationalism and the transformation of working-class politics.* Bristol: Policy Press.

Wistow, J. (2022). *Social policy, political economy and the social contract.* Policy Press.

CHAPTER 2

Capitalism's Trajectories and Local Spatial Dynamics

Abstract This chapter provides a political economic exploration of epochal change under capitalism. It begins in the mid nineteenth century and the era of laissez faire, moving into the various crises in the first phase of the twentieth century such as the 1929 Great Depression and Two World Wars. It explores how these forces combined to produce a period of relative affluence, stability and security in the post-war era. The transition to neoliberal political economy in the late 1970s set the stage for intensified deindustrialisation, the abandonment of governmental commitments to full employment, the 2008 financial crash and austerity. The chapter closes by discussing the emergence of political dissatisfaction in so called 'left behind' places.

Keywords Capitalism • Neoliberalism • Deindustrialisation • Teesside

Introduction

The rapidity of social and economic change in places like R&C in Teesside in the North-East of England is intimately connected to capitalism's historical development. Capitalism's epochal shifts have served to restructure social life, particularly from the mid nineteenth century onwards with the

L. Telford, J. Wistow, *Levelling Up the UK Economy*,
https://doi.org/10.1007/978-3-031-17507-7_2

arrival of the industrial revolution (Bew, 2017; Hobsbawm, 1975; Judt, 2010). Social crises have often been central to these changes. This includes the 1929 Great Depression and World War Two which were important contributory factors to the rise of the post-war social democratic era, as well as the global oil crisis and 1978 winter of discontent which put neoliberalism in the ascendancy. Both shifts possessed myriad implications for place. As this book documents, the problems that exist in the Government's 'left-behind' areas for Levelling Up are both long-running and deeply embedded; they cannot be solved overnight (MacKinnon et al., 2022; Martin et al., 2021). Therefore, it is important to situate our argument within a macro historical context.

This chapter covers a lot of ground rather quickly, providing a brief history of periodic political economic change whilst using R&C in Teesside as an example of a 'left behind' place. It begins by briefly sketching Teesside's expansion under the industrial revolution in light of laissez faire capitalist ideology. It then briefly outlines how structural crises combined in the early twentieth century to generate capitalism's social democratic phase, engendering a period of relative economic stability and social security and the gradual narrowing of place-based inequalities (Martin, 2021; Martin et al., 2021). Next, it discusses the shift to neoliberalism in the late 1970s including the spatial implications like the gradual emergence of both a London-centric economy and one of the most regionally unbalanced nations in the developed world (Jones, 2019; UK2070 Commission, 2020). It then highlights the austerity era as well as how debates surrounding 'left behind places' are tied to the rise of political dissatisfaction vis a vis Brexit and the collapse of the Red Wall (MacLeod & Jones, 2018; Martin et al., 2022; Rodriguez-Pose, 2018; Sandbu 2020; Telford, 2022a).

Laissez Faire

The history of localised spatial dynamics is bound to capitalisms expansion and the industrial revolution, with macro political economic restructuring historically reshaping place. Whilst there is some intellectual disagreement over when capitalism emerged (Galbraithe, 1994; Streeck, 2016), before the nineteenth century most places were agricultural economies (Galbraithe, 1994; Shildrick et al., 2012). People generally worked the land, with various hamlets constituting Teesside such as what is now its main town—Middlesbrough—possessing a mere 40 residents in 1820 (Briggs, 1963; Shildrick et al., 2012). However, two local businessmen named John Vaughan and Henry Bolckow settled in the area in the 1840s and were originally involved in various small enterprises, until they stumbled across a 'magnificent supply of workable ironstone in the near-by Cleveland Hills at Eston' (Briggs, 1963 p.250) in 1850. This discovery meant the capitalist laws of motion—competition, entrepreneurialism, market expansion and profitability (Harvey, 2005)—organised the area's economy, resulting in rapid industrialist growth that was unmatched across the world. This reordered social, cultural, and economic life—'from rural backwater to industrial powerhouse' (Shildrick et al., 2012, p. 40)—as the epoch of laissez faire capitalism took hold.

Underpinning industrialism was the ideology of laissez faire within the political economy, which is often associated with Adam Smith's the Wealth of Nations (1776). Such beliefs and ideas viewed society as governed by the invisible and natural hand of the market. The state's role is to remove or weaken obstacles to market liberty including regulations, tariffs, and taxes, and thereby enact a legal, moral, and social scaffolding to award primacy to market freedoms. Individualism, private property, competition, and inequality is inevitable and celebrated since wealth gradually trickles down to the socio-economic betterment of the working classes. Wage increases are therefore not based upon employers conceding to worker's demands for improved working conditions, but the ascendency of the market via an increase in trade and market share. This systematic liberalisation of the economy caused wide discrepancies in wealth between the poorest and richest citizens; economic hardship for the many and prosperity for the few (Hobsbawm, 1975; Judt, 2010).

Therefore, laissez faire capitalism's structural conditions were debilitating for working class people. Whilst the safety net provided through the welfare state was insufficient to mitigate the market's tendency to generate high levels of unemployment and social inequalities, employment was defined by poor pay and the absence of social and economic stability and security (Harrison, 1990; Hobsbawm, 1975). Widespread poverty, deprivation and slum conditions prevailed in many areas. Unequal and fixed gendered roles were also the norm, whereby men were predominantly the sole wage earner and women were largely domestic servants, nurturing children and looking after the family home (Briggs, 1963). Leisure opportunities particularly for working class women were restricted, while men often got intoxicated at the local pub or social club and gambled on horse racing (Bell, 1907; Hoggart 1957; Telford, 2022a).

Although the first trade unions were formed in 1871, they initially possessed little power and influence in labour markets, with men often working twelve-hour days that contained no breaks (Harrison, 1990; Hobsbawm, 1975). While most employees did not have pensions, those that did often did not live long enough to reap their rewards due to the corrosive impacts of poverty and deprivation upon one's health and life expectancy (Harrison, 1990). Characterising Teesside's local economy was often physical and backbreaking employment in heavy engineering, steel making, mining, petrochemicals, and the shipbuilding industries (Lloyd, 2013; Telford, 2022a; Warren, 2018). By 1880 Britain produced a third of the globe's steel (Harrison, 1990), and Teesside was one of the nation's leading producers as part of the region's *great steel age* (Telford, 2022a). While closely knit terraced houses clustered many of Teesside's industrial towns, chimneys, smoke, and soot filled its industrialised skyline (Briggs, 1963). This had a negative knock-on effect upon the environment and local population's health and wellbeing, not least as it is associated with respiratory illnesses like bronchitis, rhinitis, and asthma (Beynon et al., 1994). However, Briggs (1963) notes how some Victorian commentators cast the prevalence of smoke and the bright evening skyline as rather joyous, symbolising employment and the area's identity.

The 'catastrophic blows of the twentieth century' (Hobsbawm, 1975 p.43) though, impacted detrimentally upon society. Known as the 'The Great War' due to its unprecedented devastation, the First World War (1914–1918) resulted in over 30 million deaths across the world, widespread homelessness, destroyed infrastructure, hunger and starvation, as well as one of history's worst pandemics via the 1918 Spanish flu which is believed to have originated in an overcrowded USA army camp (Briggs et al., 2021). Although Teesside's industrial expansion continued via the formation of Imperial Chemicals Industry (ICI) in 1926, the subregion was further hit by the 1929 Great Depression. The absence of state intervention in the economy allowed reckless speculation on international financial markets, with stock prices falling significantly. World trade fell by around 60%, giving rise to mass unemployment particularly in Northern areas that contained a rather narrow industrialised labour market (Hudson, 2004; Lloyd, 2013; Telford, 2022a), elucidating how many of today's 'left-behind' places have struggled at specific historical junctures (De Ruyter et al., 2021). Austerity was implemented across much of the Western world to try and enable the economy to bounce back, but hopelessness, despair, and profound human suffering prevailed. As we will see, capitalist crises tend to repeat themselves (Galbraithe, 1994) and this proved to be the case after the 2008 global financial crash with austerity implemented across much of the developed world, with myriad implications for the evolution of 'left behind' places.

Indeed, it was not until World War Two that industrialised regions started to recover. Rearmament for war meant the light of their manufacturing industries was turned on again, enabling many people to return to work to fuel the war effort (Beynon et al., 1994; Lloyd, 2013). Although at one point it was feared that Nazi Germany would emerge from the war victorious, the Communist ruled Soviet Union joined ranks with the allied forces of Western Europe to defeat the fascist regime and enable victory (Galbraithe, 1994; Telford, 2022a). But the European continent was in a battered state after several turbulent decades, meaning there was 'a choice between repair and revolt' (Galbraithe, 1994 p.92) particularly as Communism offered an ideological alternative to a capitalist system that had lost its legitimacy. Such structural conditions engendered a peculiarity in capitalism's résumé involving several decades of rising living standards, a universal welfare state and full employment (Bew, 2017; Lloyd, 2013; Mitchell & Fazi, 2017; Winlow & Hall, 2013; Wistow, 2022). These issues are the subject of discussion in the next section.

A Capitalist Aberration

The horrors of the early twentieth century meant Britain had 'undergone a silent revolution' (Bew, 2017 p. 348), with the UK citizenry's sentiments shifting to the political Left by the end of World War Two. Government policies reflected a demand for social reform throughout capitalism's post-war period (1945–1979), which has been cast as the 'Keynesian', 'Fordist' and 'social democratic' period (Harvey, 2005; Mitchell & Fazi, 2017; Streeck, 2016). Symbolising the UK's political economy was a compromise between capital and labour and a governmental commitment to one nation politics, with the acknowledgement by ruling elites that capitalism could only survive if it provided security and stability for the working classes (Streeck, 2016). Such a dynamic resulted in the election of Clement Atlee's Labour Party (1945–1951) on a majority of 145 seats, branded as a political earthquake since they unexpectedly ousted Winston Churchill's Conservative Party after his successful stewardship through the war (Bew, 2017). The Labour Party promised to end the 'five giant evils' identified in the 1942 Beveridge Report: Want,

Disease, Ignorance, Squalor, and Idleness. For example, the National Insurance Act 1946 was designed to tackle social inequalities and deep-rooted poverty brought by the laissez faire era, via the construction of social insurance measures (Bew, 2017). This contributed to the birth of the modern welfare state, whereby each working age citizen paid a small weekly fee that supported various benefits including unemployment, sickness, and maternity allowance, while stipulating a minimum standard of living to be free from the five evils.

The state also played a key role in the economy often through stringent planning, regulation of markets and investment (Martin et al., 2022; Sandbu 2020), nationalising core utilities such as steel, energy, railways, telecom, electricity, and coal (Bew, 2017; Judt, 2010). The National Health Service was also created in 1948, providing healthcare for free regardless of one's social and economic standing in society. Trade unions played an important role in many labour markets, holding employers to account over exploitative working conditions particularly within industrial work. A regulatory straitjacket was imposed upon markets and business activities, while wages rose quickly and living standards improved in relative terms, to an extent that had not been witnessed before in capitalism's history (Mitchell & Fazi, 2017; Telford, 2022b; Winlow & Hall, 2006, 2013). As economic inequality diminished for the first time in capitalism's longue durée (Mitchell & Fazi, 2017; Streeck, 2016; Winlow et al., 2015; Winlow & Hall, 2013), inequalities across regions declined albeit slowly (Martin et al., 2021).

Whilst the state intervened in markets to ensure the stability and security of the working classes, they also constructed a regulatory framework that aided the post-war period's class compromise (Bew, 2017; Harvey, 2005; Judt, 2010). Central to this were successive governmental commitments to prolonged investment and full employment, driven by a 'determination to avoid a return to the horrors of the thirties, where men and machines decayed in idleness' (Judt, 2010 p.358). The Conservative Government's 1963 Hailsham report identified Teesside and the broader North-East region as key for both industrial and economic growth. The report increased investment in the North-East from £55 million in 1962/3 to £80 million in 1964/5 and included regional development grants for industrial employers to utilise for new plants and machinery (Foord et al., 1985; HMSO, 1963). While ICI opened another plant in Teesside in the 1950s, they also secured 60% of the Hailsham report's regional grants to help further substantiate its industrial bases in the subregion (Warren,

2018). Full employment was achieved in some parts of Teesside, with evidence indicating that this often-undermined employer's ability to discipline workers since they could easily acquire alternative work in the area (Telford, 2022b; Warren, 2018; Williamson, 2008). As we will see, these conditions continue to shape sentiments amongst residents in the present and possess implications for Levelling Up.

The region also led the way globally in industrial production (Telford & Wistow, 2020), creating many of the 'wonders of the industrial world' including the Sydney Harbour Bridge in Australia, Golden Gate Bridge in San Francisco, USA, as well as the Indian Railway System (Shildrick et al., 2012). As Briggs (1963 p. 242) commented, iron ore had 'crept out of the Cleveland Hills, where it has slept since the Roman days, and now, like a strong and invincible serpent, coils itself around the world'. Men in particular were proud to be known as either a steelmaker, miner, engineer and shipbuilder, with sons often following their fathers and grandfather's footsteps into industrial employment (Kotze, 2019; Lloyd, 2013, 2018; Shildrick et al., 2012; Warren, 2018; Willis, 1978). Many acquired what they believed to be a 'job for life', buttressed by a sense of relative camaraderie and collegiate workplace relations on the shopfloor (Williamson, 2008). Some industrial employers provided healthcare for their workers, as well as leisure pursuits including a social club to relax and have a drink together after their shift (Williamson, 2008). Whilst much industrial work was physical, dirty, and dangerous (Ellis, 2017; Kotze, 2019), it provided a sense of social achievement and identity as employees often felt important and part of the nation's broader economic goal of industrial growth after the horrors of the first half of the twentieth century (Telford & Lloyd, 2020; Telford & Wistow, 2020).

Evidence indicates that a sense of community was fostered in some industrialised, working-class localities. As Winlow and Hall (2006) document, people went to the same school, lived in the same area, and commenced a similar industrial job upon leaving school, sometimes at the same workplace. Although industrial jobs were highly skilled, leafy educational qualifications were not a pre-requisite for employment (Sandbu 2020; Willis, 1978). Some school friends spent their whole working lives together at one employer and experienced very similar lives, generating feelings of commonality, shared purpose, and mutual relations (Hoggart 1957; Willis, 1978; Winlow & Hall, 2006). Working class life was often repetitive and homogenous, but it provided clear life trajectories and a sense of predictability as people were grounded in their neighbourhood

(Warren, 2018). Whilst much of the industrial working class were aspirational, they primarily desired collective rather than individual social and economic advancement. This attachment to place meant life was often quite parochial, as Hoggart (1957 p.33) highlights:

> The more we look at working class life, the more we try to reach the core of working class-attitudes, the more surely does it appear that the core is a sense of the personal, the concrete, the local: it is embodied in the idea of, first, the family and, second, the neighbourhood.

As we will encounter, such attachment to place is a key part of the 'left behind' and Levelling Up debate over sixty years later (MacKinnon et al., 2022; Sandbu 2020). As the hardships of the early twentieth century faded into history, structural feelings of optimism and positivity developed in some industrialised places. Middlesbrough in Teesside was regarded as one of the most relatively prosperous places in Britain; in the 1970s it stood third behind only London and Aberdeen in terms of high wages and employment opportunities (Beynon et al., 1994; Lloyd, 2013; Shildrick et al., 2012; Warren, 2018). Whilst the industrial working classes witnessed their lives relatively improve in a way that had no historical precedent—opinion polls suggest that the hot summer of 1976 was the relatively happiest time of British peoples' lives (Viven, 2009).

However, capitalism's peculiar formation in the post-war phase did not last forever. After around two decades of stability and security, structural gales started to blow again. Inflation rose to over 20% in the 1970s in part because of global oil crises of 1973 and 1978, caused by war in the Middle East, where Britain obtained much of its energy supply from (Viven, 2009). The value of sterling also depreciated which led James Callaghan's Labour Government to borrow money from the International Monetary Fund. Essentially the Party declared that the staple features of the post-war period—rising living standards, full employment, and state intervention in the economic interests of the industrial working class—were coming to an end (Martin, 1988; Mitchell & Fazi, 2017; Wistow, 2022). Hostile and embittered relations between some employers and workers also intensified, culminating in the 1978 winter of discontent. Mass strikes were called by many trade unions regarding pay disputes, leading to some employees withdrawing their labour like Ford workers, oil tanker drivers, road hauliers, dustmen, domestic cleaners, and gravediggers. Such structural conditions fostered a sense that the Labour Party had lost control over the nation's affairs, enabling capitalism to change once again. This shift to neoliberalism had myriad implications for places like R&C in Teesside and is the crucial historical turning point in the evolution of 'left-behind' locales. As Martin et al. (2021 p.30) suggest, 'it is in terms of this systemic transformation that today's problem of left behind places, in the UK and elsewhere, needs to be situated and explained, and the immense scale of the policy challenge to level up those places must be appreciated'.

The Neoliberal Phase Shift

Neoliberalism has been subjected to scholarly debate. While some have outlined how it has been criticised for being too intellectually nebulous and unable to account for the localised specificity of neoliberalism's global rollout (Peck & Theodore, 2019; Wistow, 2022), others point to how it seeks to embed market principles including competition, radical individualism, self-preservation, and the maximisation of profitability into all aspects of social life and culture (Davidson, 2017; Harvey, 2005, 2007; MacLeavy, 2019; Streeck, 2016). This has been enabled by the construction of a global institutional framework conducive to capitalist's economic interests (Slobodian, 2018), involving how in the 1980s post-war Keynesian economists were ditched from both the World Trade Organisation which governs global trade, and the overseer of world finance the International Monetary Fund, to be replaced with neoliberal economists (Harvey, 2007; Peck & Theodore, 2019). Such institutional capture constructed a shield from democratic demands, enabling global institutions to overrule the demands of nation states and thus ensure the 'global economy was safely protected' (Slobodian, 2018 p.264) from redistributionist politics.

These institutions foist neoliberal ideology upon governments involving primacy to the four freedoms—freedom of movement for capital, labour, services, and commodities—enabling capital to move across borders to exploit less stringent tax regimes and poorly regulated labour markets (Slobodian, 2018). Neoliberal ideology also emphasises primacy to market forces instead of the needs of the citizenry, a lightly regulated, globalised marketplace involving low taxes for MNCs, the privatisation of public amenities that historically existed outside of capitalist markets such as energy, transport, healthcare, and education, as well as the flexibilization of the labour market (Harvey, 2007; Martin et al., 2022; Streeck, 2016; Winlow & Hall, 2013; Wistow, 2022). Whilst freedom under post-war capitalism was largely thought of as being free from the five wants and free to meet one's basic needs, freedom under neoliberalism was reconstituted to partially mean economic freedom for capital to heighten exploitation and enhance profit.

Following winning a second term of government at the 1983 general election, Margaret Thatcher called for a revolution but lamented the lack of revolutionaries in her party (Davidson, 2017). However, her defeat of the miners after the bitter 1984–1985 industrial dispute was a key historical moment and ushered in rapid industrial retrenchment. Within ten years of the miners' strike coming to an end, around 90% of Britain's coalmining workforce had lost their jobs (Beatty et al., 2007). In Teesside, tens of thousands of manufacturing jobs were lost across the 1980s in steel, textiles, and shipbuilding, meaning parts of Teesside like central Middlesbrough possessed a 40% unemployment rate by 1982 (Foord et al., 1985), while unemployment in the Northeast reached a record level of 24% in 1984 (Beynon et al., 1994). The rapidity of capital flight and economic decline in places like Teesside, whereby many locales had their economic heart ripped from them, means the sub-region quickly became not only one of the most deindustrialised places in Britain but arguably in Western Europe too (Beynon et al., 1994; Warren, 2018).

The surge in joblessness in the 1980s was cast by Thatcher as unfortunate but both a necessary and temporary price to pay for the epochal shift to a more globalised and competitive economy (Hall et al., 2008), though many of these locales form some of the key 'left-behind' localities for the Levelling Up agenda over three decades later (see: Martin et al., 2021). Notwithstanding, Harvey (2007 p.38) states that neoliberalism 'entailed the deliberate creation of unemployment' to create a reserve army of labour for employers to draw upon when required after capital's economic power had been restrained in capitalism's post-war phase. Chief Economic Advisor to the Treasury between 1991–1997, Sir Adrian Budd, openly admitted this by claiming raising joblessness 'was an extremely desirable way of reducing the strength of the working classes' (Cited in Davidson, 2017 p.619) and allow capitalists to yield greater profitability. Therefore, the normalisation of a level of unemployment throughout neoliberalism has been an economic gift for capitalists, undermining worker's bargaining power as they know they can be rather easily replaced (Davidson, 2017; Harvey, 2005; Streeck, 2016; Winlow & Hall, 2006).

As industries like coal, iron, steel, and chemicals were historically concentrated in the Northeast, myriad regional inequalities emerged in light of deindustrialisation (Hudson, 2004, 2013) with localities that were more economically diverse and less dependent upon manufacturing enduring a far less severe form of economic decline (Martin, 1988; Martin et al., 2021). From 1979 to 1987, job opportunities across the North declined by 1.3 million while jobs in the South increased by 66,000 (Martin, 1988). Not surprisingly a debate emerged regarding British society being characterised by a North-South divide (MacLeod & Jones, 2018; Martin, 1988). The North was defined by unemployment and social fragmentation and the South was far more socially and economically prosperous and 'pulling ahead' of the rest of the nation, cast as the 'two Britains' (Martin, 1988 p.397). Such geographical disparities were heightened by Thatcher's 'Big Bang' deregulation of financial markets in 1986, laying the foundations for the emergence of London as the nation's economic powerhouse (Martin et al., 2021). It is worth briefly considering the case of London since how it has 'pulled ahead' from the rest of Britain is a central part of the Levelling Up debate (Leyshon, 2021; Martin et al., 2021), which is also further discussed in the next chapter.

In the 1960s London was an industrialised city, possessing around 1 million employees in manufacturing work particularly in dockland areas (Rowthorn, 2010). Whilst London suffered from deindustrialisation in the 1980s, it was able to use its position as a centre for global financial capital to reinvent itself after the Big Bang, with the capital obtaining a disproportionate number of remunerative jobs in financial services and professional business (Hudson, 2013; Martin et al., 2021; Rowthorn, 2010). As we will see, particularly since the 2008 global financial crisis London has emerged as a hub for the world's super rich who largely live in an enclosed bubble of ostentatious wealth; cut adrift from civil ties and social responsibility (see: Atkinson 2020; Winlow & Hall, 2013). Nevertheless, since the 1980s politicians have repeatedly emphasised the benefits of global competition and 'letting London rip' (Leyshon, 2021 p. 1683), based upon the idea that wealth will trickle down to the rest of the nation.

Although some commentators suggest neoliberalism has withdrawn the nation state, others intimate that it has restructured the state to service capital's interests (Fine & Saad-Filho, 2017; MacLeavy, 2019; Mitchell & Fazi, 2017; Peck & Theodore, 2019; Streeck, 2016). At the national level, it thus requires political regimes of consolidation to cement its ideological principles in society (Davidson, 2017). The New Labour government (1997–2010) were a key part of this neoliberalisation process (Etherington,

2020), entrenching market values, privatising state assets and allowing social, economic, and regional inequalities to increase (Etherington, 2020; Telford & Wistow, 2020; Winlow et al., 2015; Winlow & Hall, 2013). While New Labour invested in the welfare state, reduced child poverty, and introduced the national minimum wage, they allowed the flexibilization of the labour force to grow alongside the further deindustrialisation of the British economy, heightening socio-spatial imbalances. Further industrial job loss after the turn of the millennium had debilitating impacts at the micro level in some of Britain's 'left behind' places, intensifying joblessness, mental health problems and a lost identity (Telford, 2022b; Telford & Lloyd, 2020; Warren, 2018).

Based upon a commitment to primacy to market forces and a welfare safety net to alleviate the market's tendency to generate social problems, the politics of New Labour was dressed up as the 'Third Way' (Peck & Theodore, 2019). However, it was arguably little more than a regime that embedded the neoliberalisation process (Wistow, 2022), increasing the economic insecurity and social instability of much paid employment, with power shifting away from workers towards employers (Etherington, 2020; Fine & Saad-Filho, 2017; Harvey, 2007; MacLeavy, 2019; Streeck, 2016). This is particularly the case in Britain's deindustrialised and 'left behind' locales, whereby an economic transition to service-based employment has not been a sufficient replacement for a productive economy (Martin et al., 2021). Often characterised by non-unionisation and low-pay, employment at the lower ends of these area's labour markets tends to be degrading and unable to offer the economic platform required to forge a stable livelihood (Lloyd, 2013, 2018; Winlow & Hall, 2006). Despite often working long and unsociable hours, many low-paid workers often must rely upon support from the welfare state to get by and pay the bills (Mckenzie, 2017a, 2017b). As we will encounter, in the austerity age this safety net has been punitively eroded, resulting in increased spatial imbalances and social and economic deprivation particularly in 'left behind' places.

With the erosion of trade union power, through a significant decline in trade union membership and strike action, labour markets have become increasingly 'flexible'; a watchword of neoliberalism (Kotze, 2019; Mitchell & Fazi, 2017; Streeck, 2016; Winlow & Hall, 2013). Short-term, part-time, and fixed-term work contracts have become prevalent, though most workers would take on more hours or a full-time job if they were offered one (Lloyd, 2018). Some employers increasingly use targets as a tool to discipline the workforce, since failure to meet them often results in the refusal to pay annual wage increases or bonuses (Lloyd, 2018; Telford & Briggs, 2022). As a reserve army of labour has become normalised, employers often utilise this to heighten workplace exploitation and thereby impose more targets and workplace demands, intensifying workers' stress and mental health problems (Telford & Briggs, 2022). Both employers and workers know that the latter can be easily disposed of, generating insecurity and instability.

As many previously industrialised locales including R&C in Teesside lost their raison d'être under neoliberalism, higher than national average levels of unemployment, poverty, homelessness, poor educational attainment, family breakdown and criminal activity, have followed (Martin et al., 2021; Shildrick et al., 2012; Telford, 2022b; Treadwell et al., 2020). As Hall et al. (2008 p.22) previously indicated, 'it is now clear that pockets of acute marginalisation have developed in areas of permanent recession throughout Britain', whereby crime often offers one of the economic means to forge a livelihood (Ancrum & Treadwell, 2017; Ellis, 2017). As we will encounter, drug misuse particularly of heroin, crack cocaine and alcohol, acquisitive crime including shoplifting, as well as violence is often normalised and endemic in some deindustrialised and 'left behind' places (Ancrum & Treadwell, 2017; Ellis, 2017; Hall et al., 2008; Kotze, 2019; Treadwell et al., 2020).

The neoliberal shift punctured many deindustrialised area's 'socio-economic organs' (Kotze, 2019 p.41), resulting in the emergence of a reserve pool of labour, intensification of regional inequalities particularly between the North and South and the degradation of workplace conditions. The next section explores the age of austerity. This historical phase further aggravated both socio-spatial inequalities and political dissatisfaction culminating in Brexit and the fall of the Red Wall, resulting in the emergence of debates about 'left behind' places (MacKinnon et al., 2022; MacLeavy & Jones, 2021; MacLeod & Jones, 2018; Martin, 2021; Martin et al., 2021; Peck & Theodore, 2019).

Austerity

Starting in the USA where financial markets had undergone extensive deregulation, the 2008 global financial crisis quickly spread throughout neoliberal globalisation's economic organs, generating widespread panic and fear amongst the citizenry (Streeck, 2016; Telford 2022; Winlow & Hall, 2013). Bankers had been engaging in nefarious financial practices including offering subprime mortgages to people in poverty-stricken areas on meagre incomes, coupled with histories of long-term unemployment (Telford, 2022a). Whilst these financial instruments were thus inherently risky as there was a high risk of individuals defaulting on their repayments, rating agencies awarded them AAA, signifying individuals' extremely strong capacity to repay their debts. While some commentators pathologized the crash and claimed it resulted from unethical and immoral banking practices, this ignores the context under which it occurred. Bankers are regularly awarded financial incentives like bonuses for meeting targets meaning there is a focus on short-term goals at the expense of long-term considerations, generating a culture of risk taking and excess (Fisher, 2018; Winlow & Hall, 2013). Whilst there is often a close ideological link and revolving door between the banking industry and political class, the culture of the industry hinges upon the idea that they are too big to fail (Winlow & Hall, 2013; Wistow, 2022).

Although former British Prime Minister Gordon Brown claimed in 2009 that the neoliberal era was over (Farnsworth & Irving, 2018), moving away from neoliberalism requires the presence of an alternative ideology and this was absent from politics during the crash. As Peck (2013 p.135) notes, neoliberalism was 'brazenly rebooted with more or less the same ideological and managerial software, complete with most of the bugs that caused the breakdown in the first place'. Therefore, the banking industry was bailed out to the tune of around £500 billion (Etherington, 2020), representing the biggest form of economic intervention from the state since World War Two. At the same time, neoliberalism absorbed the crisis and utilised it for its own expansion, morphing into a slightly different form involving cosmetic adjustments (MacLeavy, 2019; Peck, 2013; Winlow et al., 2015). One adjustment was the imposition of austerity, encouraged by international economic institutions such as the IMF (Farnsworth & Irving, 2018; Gray and Barford 2018) and adopted with much enthusiasm by the 2010 coalition government in the UK.

As bank bailouts transformed a private financial crisis into a national government debt crisis, most politicians suggested austerity was inevitable (Winlow et al., 2015). Whilst there was some disagreement over its scale and depth, most politicians claimed austerity was essential to reduce the deficit, balance the financial books like a household and restore economic growth (Gray and Barford 2018; Etherington, 2020; Peck, 2013). This served to depoliticize the financial crisis (Davidson, 2017; MacLeavy, 2019), allowing punitive cuts to be implemented particularly to local authorities' budgets and the public sector, branded as 'fiscal starvation' (MacLeod & Jones, 2018 p.119). As Gray and Barford (2018 p.544) outline, across 2010–2015 the Department of Local Government and Communities lost over half of its funding, serving as a form of 'fiscal discipline' and thereby curtailing the capacity of local governments to address localised inequalities. This is an issue that is explicated further in Chap. 4.

Neoliberalism entails various ideological shocks whereby key components of the previous capitalist phase are assaulted (Davidson, 2017). Reducing the welfare state has been central to this post-2008 as it was a key aspect of both the coalition government's (2010–2015) and Conservative administration's (2015–2019) austerity programmes. Based upon the notion that welfare recipients are lazy, feckless, irresponsible and need nudging into employment (Mills, 2018; Pemberton et al. 2016), some of the major welfare cuts have involved Housing Benefit, a Benefit Cap involving a new threshold of payments per household, the replacement of Disability Living Allowance with Personal Independence Payment, more testing and conditionality for receipt of Employment and Support allowance, as well as the increased usage of benefit sanctions (Beatty & Fothergill, 2018; Etherington, 2020). Such measures have been relatively easy for governments to implement as organised opposition has been significantly weakened under neoliberalism (Etherington, 2020). Thatcher's aim to not only to change the economic foundations of Britain, but to alter the citizenry's hearts and minds has had significance to this day (Peck & Theodore, 2019; Winlow et al., 2015). Evidence indicates that many people are more individualistic and less sympathetic towards welfare recipients, often suggesting they are solely responsible for their economic hardship (Pemberton et al. 2016; Shildrick et al., 2012; Winlow & Hall, 2013, Winlow et al., 2017).

While Conservative governments repeatedly indicated that 'we are all in this together', the impacts of austerity were not felt evenly, amplifying socio-spatial discrepancies (for instance: Bambra & Garthwaite, 2015; Beatty & Fothergill, 2017, 2018; Etherington, 2020; Gray and Barford 2018; Grover, 2019; Pemberton et al. 2016). As the demand for welfare is concentrated amongst certain social groups in specific localities that are generally 'left behind', across 2009–2010 to 2016–2017 cuts in areas ranged from 45% to 1.6% (Gray and Barford 2018). According to Beatty and Fothergill (2018) three types of areas suffered the most: 1) old industrial areas of Scotland, Wales, and England particularly in the Northeast, 2) many seaside towns and 3) some London boroughs. Deindustrialised locales across the North-East like Middlesbrough and Newcastle have arguably suffered the most, leading Bambra and Garthwaite (2015 p.341) to suggest that austerity impacted detrimentally on 'the older industrial areas in the North, while the South (outside London) escapes comparatively lightly'.

Whilst localities with higher rates of joblessness will inevitably be amongst the worst hit as they contain high rates of welfare recipients (Beatty & Fothergill, 2018; Gray and Barford 2018), austerity served to further discipline and atomise the working class (Etherington, 2020). This is particularly achieved through sanctions; across 2008–2014 there was a 69% increase in Jobseekers Allowance sanctions, coupled with an 84% rise in Employment and Support Allowance sanctions (Grover, 2019). Punitive fit to work assessments have also been rolled out by the government's private contractors like Maximus and Atos, whereby people with severe physical and mental health problems have been found fit to work (Grover, 2019; Mills, 2018). Suicides have been linked to this component of austerity, including a man with brain damage who hung himself after being informed that his Job Seekers Allowance was being stopped (Mills, 2018). Given such conditions, it is not surprising that there was a myriad rise in relative deprivation, food banks, mental health problems and antidepressant prescription usage in the era of austerity (Grover, 2019; Mills, 2018), particularly in 'left behind' areas of Teesside and the Northeast (Bambra & Garthwaite, 2015). At the same time, economic and regional inequalities increased, with London central to this.

Across 2010–2016 jobs in the nation's capital grew by around 800,000; four times the rate of the rest of Britain (Beatty and Fothergill 2020), while areas 'blighted by deindustrialisation' (Hudson, 2013, p.379) across the North continued to decline. The UK is now one of the most regionally unequal countries in the developed world (Beatty and Fothergill 2020; Jones, 2019; Martin, 2021; Wistow, 2022), with London becoming the global home of the world's richest members of society with large parts of the city unaffordable for most people as its property prices are astronomical (Atkinson 2020; Beatty & Fothergill, 2017). While London still contains deprived areas like Croydon and Barking and Dagenham, its prosperity is partially fuelled by emerging wealthy Boroughs such as Tower Hamlets and Camden (Martin et al., 2021). As Atkinson (2020) outlined, London's top ten boroughs via property wealth are now worth more than the property wealth of Scotland, Northern Ireland and North Wales put together, while it possesses the most 5-star hotels of any city in the world.

As many deindustrialised and 'left behind' locales in particular across the North continued to decline under austerity, it is unsurprising that political discontent grew in these places, expressed most evidently with their support for Brexit and the 2019 collapse of the Red Wall (Bromley-Davenport et al., 2019; MacLeavy & Jones, 2021; Telford & Wistow, 2020). This put place-based inequalities in the political ascendancy, resulting in 'left behind' places emerging as the UK's most important political battlegrounds (Peck & Theodore, 2019), and thereby central to the Levelling Up agenda.

'Left behind' Places

Recent political earthquakes in British politics—Brexit and the 2019 collapse of the Red Wall—resulted in the intensification of both spatial language and place-based inequalities as the subject of debate, embodied most lucidly in the notion of 'left behind' places and regions (Leyshon, 2021; MacLeavy & Jones, 2021; Sykes, 2018). Whilst this idea is not new and has often been attached to various marginalised groups including the USA's No Child Left Behind Act 2001 (Martin et al., 2021), it has intensified in usage in recent years (Gordon, 2018; Leyshon, 2021; MacKinnon et al., 2022; MacLeod & Jones, 2018; Martin et al., 2021, 2022; Mckenzie, 2017a; Sandbu 2020; Telford, 2022a). The term burst into popular consciousness with Brexit in 2016, where many places particularly across the North voted to Leave by a sizable margin (Sykes, 2018; Telford & Wistow, 2020). Although many areas in the South voted for Brexit including Cornwall (56.5%), South Gloucestershire (52.7%) and South Somerset (57.2%), the vote received large amounts of support across the deindustrialised and relatively deprived places and regions of the North (Rodriguez-Pose, 2018; Winlow et al., 2017). As Goodwin and Heath (2016) pointed out, the regions that voted most for Leave were in the North: West Midlands (59%), East Midlands (59%), Yorkshire and Humberside (58%) and the Northeast (58%), encompassing most of the old industrial locales within Tyneside and Teesside (MacLeod & Jones, 2018; Telford, 2022a; Telford & Wistow, 2020). This includes R&C who possessed a Leave vote of 66%. These post-industrial places tend to possess higher levels of pensioners, lower-skilled and less well-educated individuals, as well as:

> 'Citizens who have been pushed to the margins not only by the economic transformation of the country over recent decades but also by the values that have come to dominate a more socially liberal media and political class. In this respect the vote for Brexit was delivered by the 'left behind' (Goodwin & Heath, 2016, p. 331).

Although the aetiology of the rise of political dissatisfaction is beyond this book's scope and has been covered elsewhere (for instance: Bromley-Davenport et al., 2019; Eatwell & Goodwin, 2018; MacLeavy & Jones, 2021; Telford, 2022a; Winlow et al., 2017), it is important to outline some of the key characteristics of the 'left behind' places that voted for both Leave and the Conservative government in 2019, often for the first time in their electoral history. Whilst scholars have highlighted how the idea—'left behind'—is loosely defined—it is not clear precisely who or where is 'left behind' (Tomaney et al., 2021)—they encompass different types of locales including cities, towns, neighbourhoods, deindustrialised places, and former relatively prosperous coastal areas (Martin et al., 2022). These places tend to be characterised by relative economic decline, poorly paid jobs, below national average levels of employment, productivity, and educational qualifications, as well as higher than national average rates of poverty, social and economic marginalisation, poor health outcomes, depopulation, political disenfranchisement and poorer public services (Boswell et al., 2022; MacKinnon et al., 2022; MacLeavy & Jones, 2021; Martin et al., 2021; Mckenzie, 2017a; Tomaney et al., 2021). Particularly 'left behind' Northern areas tend to possess lower rates of high skilled and professionalised jobs, but higher rates of poorly skilled employment opportunities (Martin et al., 2021). Indeed, three of Teesside's five local authority districts are in the top 74 'left behind' local authority areas defined by the differential growth of both employment and output as identified by Martin et al. (2021). This includes R&C (2nd), Hartlepool (10th) and Middlesbrough (20th).

As outlined, these areas have not suddenly become blighted by problems; rather, they have been accumulating throughout neoliberalism (MacKinnon et al., 2022; MacLeod & Jones, 2018; Sandbu 2020). Such political dissatisfaction has thus been partially generated by deindustrialisation, the degradation of working conditions, distrust of politicians and a palpable sense that life is getting more and more difficult with each passing day for many people (Bromley-Davenport et al., 2019; Eatwell & Goodwin, 2018; Gordon, 2018; Martin et al., 2022; Sandbu 2020;

Telford, 2022a; Winlow et al., 2017). As mentioned, austerity also intensified socio-spatial inequalities, impacting detrimentally on public services and marginalised social groups, which has aggravated a sense of political and economic abandonment in some 'left behind' places (Mckenzie, 2017a, 2017b). Recent qualitative work in a 'left behind' locale outlined how the respondents felt their area possessed little social substance, with the high street populated by pound shops and boarded up stores (Boswell et al., 2022). Moreover, paralleling Winlow et al.'s (2017) research, many individuals reflected nostalgically upon the past, suggesting there was a better sense of community spirit, more remunerative work, and optimism about the future. According to Rodriguez-Pose (2018 p. 190) these 'left behind' localities have been deemed as 'places that don't matter' politically throughout neoliberalism, with their residents feeling as though they have been politically abandoned and possess no stake in the future.

Drawing upon focus groups and interviews with both workers and residents (N=20) in 'left behind' Sacriston in County Durham, Tomaney et al. (2021) demonstrate the locality's social and economic decline. Whilst the village was once characterised by economic stability provided by relatively well-paid industrial employment, in the neoliberal era joblessness and poor wages had become the norm. Nonetheless, some residents were attached to the area and did not desire to leave to live elsewhere. Such a sense of belonging has been a key part of the 'left behind' debate (Goodhart, 2017; MacKinnon et al., 2022; Martin et al., 2021; Sandbu 2020). Sandbu (2020), for instance, outlines how 'left behind' places and their residents once possessed an economics of belonging, since their local economies were characterised by industrial work and central to the nation's economic functioning. However, since the 1980s these places have become characterised by industrial job loss, lowly paid jobs, poor educational outcomes, crime, and mental health problems, generating a sense that they no longer possess an economics of belonging (Sandbu 2020).

As MacKinnon et al. (2022) highlight, 'left behind' places are primarily viewed through an economic lens, but as we will see they encompass far more than this. Scholars note how these areas tend to be more socially conservative in their cultural outlooks (Boswell et al., 2022; Goodwin & Heath, 2016); in particular, cities tend to be more ethnically diverse, educated and culturally liberal while many 'left behind' towns are less diverse, aging and more culturally conservative (Jennings & Stoker, 2018). This gives weight to Goodhart's (2017) assertion that the key divide in British politics today is the cultural antagonism between the 'Somewheres' and the 'Anywheres'. Whilst the latter are socially mobile, content to live anywhere and express liberal sentiments on cultural issues like immigration, the 'Somewheres' are more likely to live in 'left behind' spaces and are hostile to significant cultural change such as high net migration levels, take pride in both place and national identity and place importance on the family unit. As the next chapter discusses, this cultural schism is acknowledged in the LUWP, which outlines how Levelling Up seeks to restore a sense of community, local pride and belonging to 'left behind' places.

Nevertheless, the term 'left behind' has attracted criticism. Tomaney et al. (2021 p. 2) suggest it generates the potential 'stigmatising of their residents as passive, immobile and impoverished', while Mckenzie (2017a, 2017b) intimates that it is disingenuous because it glosses over the structural changes wrought in 'left behind' places. In fact, Mckenzie (2017a p. 208) asserts:

> 'The patronizing 'left behind' rhetoric actively supports this devalued identity of the deindustrialized working class. Rather than focus on and attempt to genuinely understand the structural nature of deindustrialization, of class inequality and of class prejudice, the 'left behind' rhetoric relies on the stereotypes and prejudices that the poor white working class are 'old fashioned', un-modern, have no mobility and long for the past'.

Therefore, Mckenzie (2017a) suggests that it would be more appropriate to cast these places as purposefully 'left out'. Whilst these localities once served an economic function often of industrial expansion and growth, they are no longer required and have been left to degenerate (Telford, 2022b; Winlow et al., 2017). Such an argument feeds into ideas about moving beyond debates on the North-South divide, cities vs towns and the 'left behind' to consider 'nested deprivation' (Boswell et al., 2022). According to Boswell et al. (2022 p. 2), these debates are too simplistic since a '20-minute drive from almost any point on the map, or a cursory glance at ONS census data, reveals that the U.K cannot be so neatly divided geographically into economic haves and have nots'. Spaces of nested deprivation could be considered 'left behind' places and include poorly paid employment opportunities, higher than national average rates of impoverishment and more socially conservative outlooks; yet they are far less discussed in political, academic, and social policy circles (Boswell et al., 2022). Whilst we acknowledge the validity of these criticisms, we agree with MacKinnon et al. (2022) that the terminology of 'left behind' provides a useful conceptual lens for academics to engage in intellectual contestation and empirical investigation, while as we will see, the term also speaks to the realities of some of the structural conditions in R&C.

The notion of being 'left behind' has been further elucidated by the COVID-19 pandemic. First identified in Wuhan province, China, in December 2019, the rapid spread of the pathogen across the world's continents resulted in a health crisis of such magnitude not witnessed since the 1918 Spanish Influenza pandemic (Briggs et al., 2021). This meant society witnessed two crises in a relatively short period of time—the 2008 global financial crash and the COVID-19 pandemic—that are 'only supposed to be once-in-a-century events' (Martin, 2021 p. 144). Whilst space precludes a lengthy exploration of the pandemic (see: Briggs et al., 2021), it is important to mention that the pandemic and its associated restrictions amplified pre-existing spatial discrepancies and social problems (Cross et al., 2022). Left behind spaces have endured higher rates of COVID-19 infection and mortality due to their above national average levels of poor population health, poor social infrastructure, and generally corrosive impact of poverty upon one's health, wellbeing, and life expectancy (Agrawal & Phillips, 2020; Martin et al., 2021).

The pandemic also further exposed increasing inequalities between various social groups, particularly the rich and impoverished, old, and young and different ethnic groups, combined with inequalities across regions and localities (Agrawal & Phillips, 2020; Briggs et al., 2021; Cross et al., 2022). Nonetheless, the pandemic amplified the sense 'that the North is falling behind the South, that towns and rural areas are struggling relative to cities, and that London is storming ahead of everywhere else' (Agrawal & Phillips, 2020, p. 5). Moreover, the intensification of digitalisation during the pandemic could impact detrimentally on those places that rely heavily upon retail, while caution about socialising at pre-pandemic levels will hit tourism the hardest (Agrawal & Phillips, 2020). The government also spent vast sums of money on supporting workers, jobs, and businesses to prevent total economic collapse and mass unemployment (Briggs et al., 2021; Martin, 2021), with current estimates standing between £310bn–£410bn (Brien & Keep, 2022) Martin et al. (2021) highlight that around 50% of the jobs impacted negatively by the pandemic and associated restrictions were in generally low-paid sectors like retail and construction. In comparison the nature of middle-class work meant many people were able to work remotely at home, while many poorly paid and low-skilled workers had no choice but to go out to work (Briggs et al., 2021). As mentioned, these tend to cluster more in 'left behind' places and regions.

Conclusion

This chapter explored capitalism's historical development from the nineteenth century onwards, demonstrating how the macro context filters into the meso and micro levels, reconfiguring place, the relationship between the state and citizenry, economic relations, and local economies (Martin et al., 2022; Mitchell & Fazi, 2017; Streeck, 2016). The UK political economy has witnessed several key phases of capitalist development, beginning with the industrial revolution and the arrival of manufacturing work in places like Teesside. Structural crises in the early twentieth century meant the post-World War Two era generated what Sandbu (2020) termed an economics of belonging; industrialised places were an important part of the nation's economy, remunerative work was relatively easy to obtain, and working-class people could look to the future with optimism that life would continue to improve. Regional inequalities narrowed under this capitalist regime (Martin, 2021), while economic inequality declined. Deindustrialisation and the turn to neoliberalism marked a 'phase shift' in the political economy and society more generally (Wistow, 2022) and enabled capital to re-establish its economic power (Etherington, 2020; Harvey, 2005, 2007.). Deindustrialisation, in particular, led to many 'left behind' places gradually losing (albeit with considerable spatial variation) their economic DNA (Martin et al., 2021; Sandbu 2020; Telford & Wistow, 2020; Tomaney et al., 2021).

The 2008 global financial crisis temporarily threatened the legitimacy of the neoliberal political economy but its longer-lasting legacy in the UK was a 10-year programme of austerity. This exacerbated regional and spatial inequalities, particularly as cuts to local authority budgets and the welfare state impacted most detrimentally upon those deprived Northern locales that had already witnessed severe economic restructuring and decline (Beatty & Fothergill, 2017, 2018; Gray and Barford 2018; MacLeod & Jones, 2018). Growing political dissatisfaction in many 'left behind' areas was expressed most clearly in Brexit and the 2019 collapse of the Labour Party's long held Red Wall. The COVID-19 pandemic further intensified and exposed how the UK economy is one of the most regionally unbalanced nations in the developed world (Cross et al., 2022).

The Levelling Up agenda has emerged as a response to growing dissatisfaction with place-based inequalities. However, 'research has yet to fully engage with the development problems of such 'left behind' places and address the aspirations and needs of their residents' (MacKinnon et al., 2022, p. 40). Therefore, the next chapter explores the Levelling Up agenda including debates prior to, and after, the publication of the LUWP. It also occasionally outlines data from the Directors of Regeneration. This lays the foundations for a more detailed discussion about the role of local government relative to Levelling Up in Chap. 4, as well as working-class people's views of the problems in their 'left behind' area in Chap. 5.

References

Agrawal, S., & Phillips, D. (2020). *Catching up or falling behind? Geographical inequalities in the UK and how they have changed in recent years.* Institute for Fiscal Studies.

Ancrum, C., & Treadwell, J. (2017). Beyond ghosts, gangs and good sorts: Commercial cannabis cultivation and illicit enterprise in England's disadvantaged inner cities. *Crime Media Culture, 13*(1), 69–84.

Bambra, C., & Garthwaite, K. (2015). Austerity, welfare reform and the English health divide. *Area, 47*(3), 341–343.

Beatty, C., & Fothergill, S. (2017). The impact on welfare and public finances of job loss in industrial Britain. *Regional Studies, Regional Science, 4*(1), 161–180.

Beatty, C., & Fothergill, S. (2018). Welfare reform in the UK 2010-16: Expectations, outcomes and local impacts. *Social Policy & Administration, 52*(5), 950–968.

Beatty, C., Fothergill, S., & Powell, R. (2007). Twenty years on: Has the economy of the UK coalfields recovered? *Environment and Planning A, 39*, 1654–1675.

Bell, L. (1907). *At the works: A study of a manufacturing town.* E Arnold.

Bew, J. (2017). *Citizen Clem: A biography of Attlee.* Riverrun.

Beynon, H., Hudson, R., & Sadler, D. (1994). *A place called Teesside: A locality in a global economy.* Edinburgh University Press.

Boswell, J., Denham, J., Furlong, J., Killick, A., Ndugga, P., Rek, B., Ryan, M., & Shipp, J. (2022). Place-based politics and nested deprivation in the UK: Beyond cities-towns, 'two Englands' and the 'left behind'. *Journal of Representative Democracy, 58*(2), 169–190.

Brien, P., & Keep, M. (2022). *Public spending during the Covid-19 pandemic.* House of Commons.

Briggs, A. (1963). *Victorian cities: A brilliant and absorbing history of their development.* Odhams Press.

Briggs, D., Telford, L., Lloyd, A., Ellis, A., & Kotze, J. (2021). *Lockdown: Social harm in the Covid-19 era*. Palgrave Macmillan.
Bromley-Davenport, H., MacLeavy, J., & Manley, D. (2019). Brexit in Sunderland: The production of difference and division in the UK referendum on European Union membership. *EPC: Politics and Space, 37*(5), 795–812.
Cross, K., Evans, J., MacLeavy, J., & Manley, D. (2022). Analysing the socio-economic impacts of COVID-19: A new regional geography or pandemic enhanced inequalities? *Regional Studies, Regional Science, 9*(1), 461–485.
Davidson, N. (2017). Crisis neoliberalism and regimes of permanent exception. *Critical Sociology, 43*(4–5), 615–634.
De Ruyter, A., Martin, R., & Tyler, P. (2021). Editorial: Geographies of discontent: Sources, manifestations and consequences. *Cambridge Journal of Regions, Economy and Society, 14*(3), 381–393.
Eatwell, R., & Goodwin, M. (2018). *National Populism: The revolt against Liberal democracy*. Pelican Books.
Ellis, T. (2017). *Men, masculinities and violence*. Routledge.
Etherington, D. (2020). *Austerity, welfare and work: Exploring politics, geographies, and inequalities*. Policy Press.
Farnsworth, K., & Irving, Z. (2018). Austerity: Neoliberal dreams come true? *Critical Social Policy, 38*(3), 461–481.
Fine, B., & Saad-Filho, A. (2017). Thirteen things you need to know about neoliberalism. *Critical Sociology, 43*(4–5), 685–706.
Fisher, M. (2018). *K punk: The collected and unpublished writings of mark Fisher*. Repeater Books.
Foord, J., Robinson, F., & Sadler, D. (1985). *The Quiet Revolution: Social & Economic Change on Teesside 1965–1985*. A Special Report for BBC North-East.
Galbraithe, J. K. (1994). *The world economy since the wars: A personal view*. Reed Consumer Books Ltd.
Goodhart, D. (2017). *The road to somewhere: The new tribes shaping British politics*. Penguin.
Goodwin, M., & Heath, O. (2016). The 2016 referendum, Brexit and the left behind: An aggregate-level analysis of the result. *The Political Quarterly, 87*(3), 323–332.
Gordon, I. (2018). In what sense left behind by globalisation? Looking for a less reductionist geography of the populist surge in Europe. *Cambridge Journal of Regions, Economy and Society, 11*(1), 95–113.
Grover, C. (2019). Violent proletarianization: Social murder, the reserve army of labour and social security 'austerity' in Britain. *Critical Social Policy, 39*(3), 335–355.
Hall, S., Winlow, S., & Ancrum, C. (2008). *Criminal identities and consumer culture*. Willan Publishing.
Harrison, J. F. C. (1990). *Late Victorian Britain: 1875–1901*. Hammersmith.

Harvey, D. (2005). *A brief history of neoliberalism*. Oxford University Press.
Harvey, D. (2007). Neoliberalism as creative destruction. *The Annals of the American Academy of Political and Social Science, 610*, 22–47.
HMSO. (1963). *The north east: A programme for regional development and growth*. HMSO.
Hobsbawm, E. (1975). *The age of capital 1848–1875*. Abacus.
Hudson, R. (2004). Rethinking change in old industrial regions: Reflecting on the experiences of north East England. *Environment and Planning A, 37*(4), 581–596.
Hudson, R. (2013). Thatcherism and its geographical legacies: The new map of socio-spatial inequality in the divided kingdom. *The Geographic Journal, 179*(4), 377–381.
Jennings, G., & Stoker, G. (2018). The divergent dynamics of cities and towns: Geographical polarisation and Brexit. *The Political Quarterly, 90*(2), 155–166.
Jones, M. (2019). *Cities and regions in crisis: The political economy of sub-National Economic Development*. Edward Elgar Publishing.
Judt, T. (2010). *Post war: A history of Europe since 1945*. Vintage Books.
Kotze, J. (2019). *The myth of the crime decline: Exploring change and continuity in crime and harm*. Routledge.
Leyshon, A. (2021). Economic geography I: Uneven development, 'left behind places' and 'levelling up' in a time of crisis. *Progress in Human Geography, 45*(6), 1678–1691.
Lloyd, A. (2013). *Labour markets and identity on the post-industrial assembly line*. Routledge.
Lloyd, A. (2018). *The harms of work: An ultra-realist account of the service economy*. Policy Press.
MacKinnon, D., Kempton, L., O'Brien, P., Ormerod, E., Pike, A., & Tomaney, J. (2022). Reframing urban and regional 'development' for 'left behind' places. *Cambridge Journal of Regions, Economy, and Society, 15*, 39–56.
MacLeavy, J. (2019). Neoliberalism and the new political crisis in the west. *Ephemera, 19*(3), 627–640.
MacLeavy, J., & Jones, M. (2021). Brexit as Britain in decline and its crises (revisited). *The Political Quarterly, 92*(3), 444–452.
MacLeod, G., & Jones, M. (2018). Explaining 'Brexit capital': Uneven development and the austerity state. *Space and Polity, 22*(2), 111–136.
Martin, R. (1988). The political economy of Britain's north-south divide. *The Royal Geographical Society, 13*(4), 389–418.
Martin, R. (2021). Rebuilding the economy from the Covid crisis: Time to rethink regional studies? *Regional Studies, Regional Science, 8*(1), 143–161.
Martin, R., Gardiner, B., Pike, A., Sunley, P., & Tyler, P. (2021). *Levelling up left behind places: The scale and nature of the economic and policy challenge*. Routledge.

Martin, R., Martinelli, F., & Clifton, J. (2022). Editorial: Rethinking spatial policy in an era of multiple crises. *Cambridge Journal of Regions, Economy and Society, 15*, 3–21.
Mckenzie, L. (2017a). 'It's not ideal': Reconsidering 'anger' and apathy' in the Brexit vote among an invisible working class. *Competition & Change, 21*(3), 199–210.
Mckenzie, L. (2017b). The class politics of prejudice: Brexit and the land of no-hope and glory. *The British Journal of Sociology, 68*(S1), 255–280.
Mills, C. (2018). 'Dead people don't claim': A psychopolitical autopsy of UK austerity suicides. *Critical Social Policy, 38*(2), 302–322.
Mitchell, W., & Fazi, T. (2017). *Reclaiming the state: A progressive vision of sovereignty for a post-neoliberal world*. Pluto Press.
Peck, J. (2013). Explaining (with) neoliberalism. *Territory, Politics, Governance, 1*(2), 132–157.
Peck, J., & Theodore, N. (2019). Still Neoliberalism? *The South Atlantic Quarterly, 118*(2), 245–265.
Rodriguez-Pose, A. (2018). The revenge of the places that don't matter. *Cambridge Journal of Regions, Economy and Society, 11*(1), 189–209.
Rowthorn, R. (2010). Combined and uneven development: Reflections on the north-south divide. *Spatial Economic Analysis, 5*(4), 363–388.
Shildrick, T., MacDonald, R., Webster, C., & Garthwaite, K. (2012). *Poverty and insecurity: Life in low-pay no-pay Britain*. Policy Press.
Slobodian, Q. (2018). *Globalists: The end of empire and the birth of neoliberalism*. Harvard University Press.
Smith, A. (1776). *An inquiry into the nature and causes of the wealth of nations*. Strahan & Cadell.
Streeck, W. G. (2016). *How will capitalism end?* Verso.
Sykes, O. (2018). Post-geography worlds, new dominions, left behind regions, and 'other' places: Unpacking some spatial imaginaries of the UK's 'Brexit' debate. *Space and Polity, 22*(2), 137–161.
Telford, L. (2022a). *English nationalism and its ghost towns*. Routledge.
Telford, L. (2022b). "There is nothing there": Deindustrialization and loss in a coastal town. *Competition & Change, 26*(2), 197–214.
Telford, L., & Briggs, D. (2022). Targets and overwork: Neoliberalism and the maximisation of profitability from the workplace. *Capital & Class, 46*(1), 59–76.
Telford, L., & Lloyd, A. (2020). From 'infant Hercules' to 'ghost town': Industrial collapse and social harm in Teesside. *Critical Criminology, 28*, 595–611.
Telford, L., & Wistow, J. (2020). Brexit and the working class on Teesside: Moving beyond reductionism. *Capital & Class, 44*(4), 553–572.
Tomaney, J., Natarajan, L., & Sutcliffe-Braithwaite, F. (2021). *Sacriston: Towards a deeper understanding of place*. University College London.

Treadwell, J., Ancrum, C., & Kelly, C. (2020). Taxing times: Inter-criminal victimization and drug robbery amongst the English professional criminal milieu. *Deviant Behavior, 41*(1), 57–69.
Viven, R. (2009). *Thatcher's Britain: The politics and social upheaval of the 1980s.* Simon & Schuster Ltd.
Warren, J. (2018). *Industrial Teesside: Lives and legacies.* Palgrave Macmillan.
Williamson, M. (2008). *Life at the ICI: Memories of working at ICI Billingham.* Teesside Industrial Memories Project.
Willis, P. (1978). *Learning to labour: How working-class kids get working class jobs.* Routledge.
Winlow, S., & Hall, S. (2006). *Violent night: Urban leisure and contemporary culture.* Berg Publishers.
Winlow, S., & Hall, S. (2013). *Rethinking social exclusion: The end of the social?* SAGE.
Winlow, S., Hall, S., Treadwell, J., & Briggs, D. (2015). *Riots and political protest.* Routledge.
Winlow, S., Hall, S., & Treadwell, J. (2017). *The rise of the right: English nationalism and the transformation of working-class politics.* Policy Press.
Wistow, J. (2022). *Social policy, political economy, and the social contract.* Policy Press.

CHAPTER 3

The Levelling Up Agenda

Abstract This chapter explores the UK Government's Levelling Up strategy and how it has proposed to remedy place-based inequalities. It explores debates prior to the publication of the Levelling Up White Paper in February 2022 including how the strategy was criticised for being ideologically ambiguous, poorly conceptualised and suffering from a lack of aims and objectives. It reiterates the highly geographically unequal and spatially imbalanced context in which Levelling Up is operating within. The chapter discusses the White Paper's approach to systems reform, its focus on capitals and the missions and policy programme that it sets out. Findings from a small-scale scoping study with Directors of Regeneration are also employed to shed light on how the policy programme is being interpreted.

Keywords Levelling Up • Spatial imbalances • Inequalities

Introduction

The Levelling Up agenda was launched during the 2019 general election campaign, in response to some of the deep-rooted problems particularly in 'left behind' places outlined in the previous chapter. As a flagship government policy, it has received considerable attention and criticism for being rather ambiguous and poorly conceptualised, particularly through the

L. Telford, J. Wistow, *Levelling Up the UK Economy*,
https://doi.org/10.1007/978-3-031-17507-7_3

absence of measures of success. There is a narrow focus upon physical infrastructure investments and not developing a fully integrated understanding of the needs and priorities in so-called 'left behind places' like R&C (Carr-West & Sillett, 2021; Connolly et al., 2021; House of Commons Business, Energy and Industrial Strategy Committee, 2021; Martin et al., 2021). In February 2022, a much debated and delayed White Paper was published by the government, which will be a key focus of this chapter. The chapter is divided into two parts. The first part deals with the period following the 2019 general election through to the production of the White Paper. Here we consider policy development and rhetoric under the umbrella of Levelling Up, alongside critiques from academia and practitioners about the strategy. In particular, the criticism of the strategy prior to the White Paper will be summarised across some broadly consistent themes. The second part of the chapter focuses on the White Paper in terms of how it understands and conceptualises problems of spatial disparities in the UK and what the policy responses to this are. We will explore systems, capitals and the missions and policy programme of the LUWP.

Levelling up: Prior to the White Paper

"It's a fancy title, though I'm not sure it is particularly meaningful. Levelling Up is trying to please too many masters" Director of Regeneration, Northern 'Other City.'

Levelling Up was the signature policy of the 2019 Boris Johnson government and despite speaking about a clear need to address huge variations in wealth, health, life chances and well-being in the country, fundamental questions remained around the agenda two years into the administration (Carr-West & Sillett, 2021; Jennings et al., 2021; Martin et al., 2021; Telford, 2022; Tomaney & Pike, 2020). The rhetoric surrounding the Levelling Up agenda following its launch and through to the White Paper could be described as a catch-all policy that possessed limited substance. Even websites like Conservative Home stated in January 2020 that nobody knows what Levelling Up means (Tomaney & Pike, 2020). The agenda has variously been described as incoherent, internally inconsistent, lacking an ideological anchor, and unlikely to lead to the transformative change it seeks (Jennings et al., 2021; Martin et al., 2021; Tomaney & Pike, 2020). A key factor in the development of the strategy has been the 'politics of levelling up', which Jennings et al. (2021) describe as the Conservative Party responding to a 'geography of discontent' stemming from uneven patterns of economic and social development, which were explored in the previous chapter. Tomaney and Pike (2020) continue to argue that the complexities of neoliberal globalisation and technological change make the loss of status and neglect of place far from straightforward issues to address, which require long-term thinking and sustained follow-through. Nevertheless, Leyshon (2021) highlights some cause for optimism given that the agenda has made long-term uneven spatial development and economic geography a much stronger feature of policy formulation in the UK, albeit with the caveat that it requires persistent attention. In this respect, the government had identified that there is a serious problem to address; but had only taken tentative steps towards this prior to the launch of the LUWP.

In July 2021, the then Prime Minister—Boris Johnson—gave a Levelling Up speech called 'Plan for Growth', arguing that Levelling Up was about more than improving people's incomes and should include health outcomes, educational opportunities, liveability, and improved infrastructure (Shearer et al., 2021). However, what Levelling Up meant in practice remained unclear with Shearer et al. (2021) citing examples of whether it concerned inequalities between people or places; what role regional cities would play; and whether the government wants to decentralise power. At this time, the House of Commons Business, Energy and Industrial Strategy Committee (2021, p. 3) were concerned whether the 'lack of definition will result in failure to deliver meaningful change for people across the country.' Similarly, Connolly et al. (2021, p. 9) described the agenda as a 'centralised model masquerading as localism', which has done little to foster local leaders' capacity to address the inequalities of 'left behind' places. Given the above, it is difficult to disagree with Jennings et al. (2021, p. 303) that the agenda represents a tendency to 'govern through political spectacle' via 'a few big infrastructure projects in the North of England, a scattering of free ports and gigafactories, refreshed high streets, or an exodus of civil servants to the regions.' Shearer et al. (2021, p. 7) usefully summarises the main policies prior to the LUWP in more detail below:

- Levelling Up Fund—£4.8bn fund of 'shovel-ready' infrastructure projects starting with a first round of funding submitted between March and June 2021 and running through to 2024/25.

- Towns Fund—£3.6bn fund for towns in England. A total of 101 towns were eligible to bid for up to £25m with investment focusing on: urban regeneration, planning and land use; skills and enterprise infrastructure; and connectivity.
- Community Renewal Fund—£220m fund providing a bridge between EU structural funds and the planned Shared Prosperity Fund. Largely revenue funding for: skills; supporting people into employment; local business; and communities and place.
- Freeports—eight locations (East Midlands Airport, Felixstowe and Harwich, Humber region, Liverpool City Region, Plymouth, Solent, Thames, as well as in Teesside on the border of R&C which many residents in Chapter Five's case study mentioned) with low-tax and low-tariff business zones.
- Skills Fund—£2.5bn fund for adult skills.
- UK Infrastructure Bank—government owned bank that will have £12bn of capital for lending and investment and can issue up to £10bn of guarantees.
- Civil service relocation—a target to move 22,000 jobs away from London and the south-east by 2030.

It is notable that the first four policies are all centrally managed and require places bidding for the allocation of funds (Shearer et al., 2021). For the biggest pot of money HM Government (2021) produced a 'Levelling Up Fund: Prospectus' that sets out a competitive funding model to be invested through local authorities in projects up to £20million that a range of tiers of local authorities, combined authorities, mayoral authorities, and the Greater London Assembly can bid for. There are a variety of complex local and sub-regional governance systems at play here and the coordinating mechanism is to add another tier—of not necessarily coterminous—administrative boundaries through the expectation that the local MP will prioritise one bid for their constituency. Connolly et al. (2021) argue that the role of MPs in championing local bids is concerning given the differential access they have to government. More fundamentally, Carr-West and Sillett (2021) identified the lack of transparency as a further cause for concern, arguing the funding process was open to accusations of bias. They also support a change in approach, citing a House of Lords Committee recommendation that the Index of Multiple Deprivation (IMD) is used when making Levelling Up funding decisions. The House of Commons Business, Energy and Industrial Strategy

Committee (2021) similarly highlighted concerns about the centrally controlled competitive funding scheme but stopped short of proposing a move away from this model and instead argued that clear funding metrics for delivery and increased resourcing to enhance capacity to bid for funding should be put in place.

However, given that the method for calculating the index has been criticised for not considering deprivation within and between areas more generally, the House of Lords approach is a better solution from a redistributive perspective. Nonetheless, as Connolly et al. (2021, p. 526) argue 'all the cards are held by government in terms of funding decisions', and they do not have plans to change the design of funding allocation. All the Directors of Regeneration we spoke to (notably most strongly in the 'Other City', 'Medium Town' and 'Large Town'—that were ranked most 'left behind' on the Martin et al. (2021) classification) were critical of the funding model and the allocation process. We will return to issues with funding later in this chapter and in Chapter Four. For now, a brief summary of the criticisms of the model from the Directors of Regeneration include:

- Capacity to bid for funds remains an issue and the process is intensive and time-consuming, often involving unrealistic deadlines, given available capacity and competing local priorities/commitments.
- Deprivation was a significant issue (resulting from and shaping the 'left behind' problem) in the localities but did not feature strongly enough in the allocation of funds.
- The funding schemes and successful bids were not considered to be well connected as a whole at both a regional and sub-regional level (with the partial exception of some transport funding).
- Intra-combined authority working could be beneficial but central government was reluctant to 'step-in' to deal with the allocation of shared funding, such as the benefits of freeports not being distributed relative to deprivation across combined authorities.
- The competitive model of funding for capital-based projects led to one-off visible projects that failed to address deprivation directly.

One of the participants ('Northern Other City') called the formula for calculating need for the Shared Prosperity Fund a "bastardised version of the IMD" because it did not consider low household income or low employment numbers. Therefore, they challenged:

"The concept of 'levelling up' when it isn't actually weighted to reflect the levels of deprivation and, as you know, there is a direct read across to the health status and mental health status in communities and ultimately things like crime stats."

This is an example of the challenges of Levelling Up when faced with the spatial disparities resulting from being a 'left behind place', which will be explored in more detail regarding R&C in Chapter Five. Moreover, there was a degree of resignation about the overall approach to funding from the Directors in the three most 'left behind places' largely because of this failure to conceptualise need and allocation around deprivation.The Director of Regeneration for the 'Midlands Northern Town', for instance, was not optimistic that funding would be reoriented towards need. In this respect, Jennings et al. (2021, p. 303) warn that 'Levelling Up may succeed through redefining redistribution to be more about status, recognition and standing rather than resources or equitable outcomes.' We will see below that the LUWP puts more meat on the bones in terms of outcomes; but the concern about funding and the implications this has for leading to more equitable outcomes remains. Furthermore, if redistribution is redefined or moves closer to being understood as symbols associated with a narrative of success (Jennings et al., 2021), it lowers the bar for redistribution and weakens the concept. However, it was clear from our interviews with the Directors of Regeneration that they were not convinced about this conceptualisation, principally because of the poor alignment of the funding model and mechanisms provided through the agenda. We will discuss these further, relative to the role of local government in the Levelling Up agenda, in Chapter Four.

To conclude this section, below is a brief summary of the consistent criticisms of the Levelling Up agenda prior to the publication of the LUWP:

- A lack of clear goals and/or clear measures for success and failure (see: Carr-West & Sillett, 2021; Connolly et al., 2021; House of Commons Business, Energy and Industrial Strategy Committee, 2021 and Shearer et al., 2021)
- Limitations of, and inequities within, the funding model (see: Connolly et al., 2021 and House of Commons Business, Energy and Industrial Strategy Committee, 2021). This includes accusations of gerrymandering, cronyism and pork-barrel politics associated with the allocation of funds (Financial Times, 2021; Good Law Project, 2021).

- A lack of trust in, and resourcing and powers for, local leaders to drive the Levelling Up agenda (Connolly et al., 2021; UK2070 Commission, 2020).
- A lack of clarity about devolution, in general, and for which tiers of devolved and local government are responsible for delivering which Levelling Up objectives (House of Commons Business, Energy and Industrial Strategy Committee, 2021; Jennings et al., 2021).
- A lack of attention to the impact of austerity on 'left behind places' and for Levelling Up (Jennings et al., 2021; Marmot, 2022).
- Strategic policy fuzziness and a lack of appreciation of network complexity (Connolly et al., 2021), including the absence of a theory about whether, and how, to connect urban areas (Jennings et al., 2021).
- A lack of clarity about how the people-based and place-based elements of Levelling Up interact (Shearer et al., 2021).

We have dealt with some of these above and return to them in more detail through the policy-related discussion and analysis in Chapter Four. It is also important to recognise that the above is set within the context of path dependencies and trajectory of a wider political economy systems approach set out in Fig. 1.1 in the introduction and explored in Chapter Two. In this respect, Martin et al. (2021) provides a very useful reminder that recent discourse in the UK has focused on an insufficiently 'place-based policy' approach, given that over 90 years of local, regional and urban policies have witnessed geographical inequalities not only persisting but intensifying in recent years. However, they (2021, p. 88) argue that place-based policy is unlikely to be the panacea some claim it may be necessary, but it is unlikely to be sufficient. The early signs emerging from the Levelling Up agenda prior to the White Paper was that as a place-focused policy it had significant areas for improvement given the heightened expectations the rhetoric raised.[1] The LUWP provided an opportunity to consider and begin to reconcile the nascent agenda, which we explore next in relation to a complex systems-based understanding.

[1] Albeit the ways in which these expectations have been raised are somewhat confused and ambiguous and are met with varying degrees of cynicism and resignation among participants in our study.

The Levelling up White Paper

As Liddle et al. (2022, p. 2) argue, the White Paper moves the agenda on from a nebulous slogan but still has limited detail on 'how the various targets, objectives and 12 missions are to be achieved and policy coordinated and delivered at spatial levels.' We agree and begin our analysis through a critical discussion of the systems approach adopted through the White Paper's concern with systems reform before exploring the approach to capitals developed in the LUWP. The section concludes with a discussion about the missions and policy programmes set out in the LUWP.

Systems Reform

The Levelling Up policy programme has been set within a complex system framing. The LUWP develops a systems approach in response to historic criticisms about spatial policy in the UK and also in response to criticisms of the Levelling Up agenda in the run up to the publication of the White Paper. In the introduction chapter, we outlined an approach to understanding complex systems and were quite explicit in tying the broader political economy into this framing, which is employed here to analyse the systems approach in the White Paper. There are three main areas that the analysis of the systems framing is organised around: what is being levelled up to and what is being 'left behind'; the relationship between the political economy and the gearing of complex systems; and implications of the nature of the political economy for Levelling Up.

Having developed a historic overview of the 'scale and source of the UK's geographic disparities' in Chapter One of the White Paper, HM Government (2022, p. 105) states that 'there is no simple or singular solution to reversing spatial disparities because local economies are complex systems, shaped by cumulative and interconnected economic, social and institutional factors.' The White Paper (HM Government, 2022, pp. 110–114) focuses on five lessons from past policy approaches as a key component of the chapter on 'systems reform'. These are (HM Government, 2022, pp. 117–157) linked to five pillars that underpin the policy regime, which are summarised in Table 3.1. Despite responding to some of the criticisms of the agenda through identifying important lessons from past policy and establishing pillars for policy development, the lessons are somewhat narrowly framed, and the pillars are fragile at best from a system understanding that is calibrated to both understanding and

Table 3.1 Lessons from past policy and five pillars underpinning the policy regime (HM Government, 2022)

Lessons from past policy	*Five pillars underpinning the policy regime*
Lack of longevity and policy sufficiency—The UK has been characterised by 'endemic policy churn' leading to fragmentation, hindering local capacity, capability and civic institutions, compared with more successful examples of subnational policy internationally.	**Medium term missions**—Seeks to address complex, long-term societal challenges through a mission-based approach using measurable and time-bound objectives.
Lack of policy and delivery coordination—Resulting from centralised and siloed policy has contributed to a 'low growth trap' in areas with weaknesses in the six capitals identified in the white paper.	**Reshaping central government decision-making**—Improved transparency about place-based spending; hardwire spatial considerations into decision-making and evaluation; improve coordination of central government policies at the local level; and greater focus on local places.
Lack of local empowerment—e.g., over centralisation has under-utilised local knowledge with negative consequences for spatial disparities.	**Empowering local decision-making**—a new framework for devolution and increased support for private sector partnerships to build economic clusters.
Lack of evidence, monitoring and evaluation—Given the complexity of spatial disparities more and better comparable data is needed to enhance transparency and reduce inefficiency and duplication. Also, poor institutional memory in central government about what has and has not worked.	**Role of data, monitoring and evaluation**—Focus on improving more timely, granular and harmonised data through a subnational data strategy; making this publicly available; improving data visualisation to improve decision-making; and work with local leaders to learn what works.
Lack of transparency and accountability—Little effective oversight of policies on spatial disparities.	**Transparency and accountability**—Statutory obligation to report on progress on the missions; and establish a levelling up advisory council.

integrating historical trajectory and political economy. The latter is not surprising given the ideological standpoint of the Conservative government; but the former is much more so given that the first chapter of the White Paper spends considerable time outlining 'drivers' of the UK's geographical disparities. It is the case that each of the lessons has a longitudinal dimension, but a fully calibrated system focus would consider the path dependent nature and extent of economic and social geographic disparities in the UK much more explicitly relative to systems reform. For example, in relation to a 'lack of longevity' policy churn remains a problem in the system, but it should be considered alongside (i.e., given it has co-evolved with) the path dependencies resulting from the UK's management of its transition to a post-industrial economy and society. Without addressing key features of the macro and its influence across systems it is likely that systems reform, as set out in the LUWP, will reproduce key elements of spatial disparities in the UK.

Any complex systems focus that seeks to Level Up must consider what is being levelled up to. As mentioned in Chapter Two, at a regional level of aggregation London and the Southeast have pulled away from the other regions. Alongside the Big Bang deregulation of the economy in the 1980s, a contributory factor to this trajectory has been that recoverability over the past four recessions tended to decline with distance from London, with the nation's capital proving to be remarkably resilient to the 2008 financial crisis (Martin et al., 2021). It is worth reiterating that London is the location of considerable amounts of poverty and internal spatial disparities, with house prices in particular inflating wealth disparities and pushing private renters into poverty (Argawal & Phillips, 2020; Atkinson, 2020; Travers et al., 2016). State-led gentrification has also contributed to the displacement of poorer residents for affluent newcomers and rent seekers (Broughton, 2019). However, London's position in the global economy as one of the main financial centres and as the location of the most centralised political and economic system in the OECD (UK2070, 2020), is a key component of the 'left behind' problem and of the solutions provided through the Levelling Up agenda. Martin (2015) concludes spatial economic imbalances have become institutionalised as recursive circular and cumulative forces in a national political economy concentrated in and controlled from London. From a systems perspective, London's position as a global centre for financial capitalism, wealth extraction and rentierism is significant for the system logic that functions to varying degrees across the nexus in the UK and the spatial disparities that are endemic within it.

The LUWP (2022 p. 88) describes London as 'looking quite different from the rest of the UK' and functioning through a circular agglomeration process which concentrates wealth, high skilled jobs and workers, while placing excessive demands on infrastructure and housing. It (HM Government, 2022) also highlights the benefits of the global economy and Big Bang of the 1980s for London and the Southeast in comparison to the *deep and lasting scars* former industrial areas and many coastal communities have experienced. The White Paper identifies a problem associated with the make-up of the political economy here but neither the chapter on systems nor the section on capitals (see below) come close to fully developing the implications of this for system reform either through addressing the extent of the imbalance or, more importantly, questioning the nature of the system logic flowing from a very particular mode of neoliberal and rentier capitalism centred in the capital. In this respect, Wistow (2022) highlights the selective approach to neoliberalism and the implications for the nature of spatial inequalities, contrasting the generally non-interventionist approach to de-industrialisation with the vast sums of public money used to bail out the financial sector in light of the 2008 global financial crash and the much higher spend per head on 'economic affairs' in London than other regions.

This selective approach to market-based economics has close ties to the notions of agglomeration and 'trickle down'. In discussion with the Director of Regeneration in the 'Northern Large Town', they highlighted that:

"There's been generally a policy assumption that if the capital succeeds, then the rest of the country benefits from that success. Equally, there's a presumption that if the city region is successful then the whole region secures those benefits. That is not the evidence we are seeing in 'Large Town'."

The 'left behind' problem is more complex than a high-level policy assumption derived from the New Economic Geography (NEG) and the view that spatial agglomeration is a natural market-driven process and spatial imbalances are an equilibrium outcome (Martin, 2015). Martin et al. (2021) highlights differential employment growth between 1980–2018 across regional and rural-urban processes. The North has grown below average, with cities and towns doing especially badly and the South performing much better, particularly in towns and villages/rural areas. Martin et al. (2021, pp. 63-4) argues that 'competitiveness effects' have become more significant recently and 'those places with better quality residential environments, better connectivity and increasing populations appear to have markedly stronger employment dynamism.' The LUWP (HM Government, 2022) and Levelling Up Fund both seek to address these types of issues through a focus on connectivity and infrastructure. However, whether this is sufficiently ambitious and systematic to Level Up is debatable. All Directors of Regeneration taking part in the scoping study emphasised the importance of connectivity for economic and spatial disparities; but questioned both the degree of control local actors had over this and whether the resourcing and policy levers provided through the White Paper would address local systemic weaknesses.

A key problem in the system logic of spatial policies, including in the Levelling Up agenda, is 'a refusal to acknowledge that the socio-economic context of relative deprivation, weak infrastructure and relatively poor education and training performance within England's poorest regions are hardly fertile grounds for such policies' (Copeland & Diamond, 2022, p. 12). The widening of spatial inequalities throughout neoliberalism suggests past policies have had limited impact (Martin et al., 2021). In short, a change in kind in the systems is required to make reducing spatial inequalities more tractable to spatial policies. Having focused above on the path dependent relationship between what is being 'levelled up to', its influence on the system and some key factors in the 'left behind' problem, we can begin to explore dynamics of the political economy further and how they relate to the gearing of complex systems.

Given the argument above, as mentioned in Chapter Two, it is no surprise that the COVID-19 pandemic[2] has not been a leveller with deprived and 'left behind' communities experiencing the worst consequences and higher-performing economies tending to fare better (Cross et al., 2022; UK2070 Commission, 2020). Furthermore, Martin et al. (2021, p. 83) state that to argue northern urban areas particularly cities and satellite towns 'will rapidly bounce back, borne by a returning wave of urban and big city resurgence, misreads the post-industrial trends seen over the past 40 years and effectively represents somewhat wishful thinking.' In this context, the UK2070 Commission (2020) identifies a weakness and lack of resilience in the UK's highly centralised political and economic systems. They (2020) identify four potential future scenarios reflecting a range of economic assumptions and policy interventions around 'low growth', 'dynamic recovery' and 'business as usual' or a 'convergent economy' (where levels of productivity, skills and jobs would converge towards those of London and the South). In a low growth economy, business as usual will likely lead to 'continued regional recession'. In a dynamic recovery,

[2] As briefly alluded to in Chapter Two, COVID-19 acted as a shock to the neoliberal system and has been highly significant in relation to social and economic outcomes (see Cross et al., 2022). However, it appears to not have changed the inherently inconsistent application of broadly neoliberal and rentier interests influencing the political economy.

business as usual will likely lead to 'persistent imbalances' reinforcing regional discrepancies in growth, jobs, inequalities, and deprivation, whereas a convergent economy could result in a 'dynamic recovery' through a clear shift in approach leading to regional growth.

For the UK2070 Commission (2020), the model of the convergent economy must be made the central goal of public policy. They argue that this would require £375billion investment over 25-years and include a full devolution settlement for local leaders. These proposals can be interpreted as being of the scale and type to have the potential to lead to a 'phase shift' in the political economy and of the complex systems cutting across it through redistributing and investing resources and capacity to localities with powers and local discretion to utilise these. As we shall see, the LUWP falls short of this ambition for spatial policy and the economic signs, at the time of writing in Summer 2022, fall within the low growth forecast. From a systems perspective the failure to match the UK2070 Commissions' (2020) ambitions, or create an equivalent programme, is problematic because it means that the government does not get close to changing the gearing of the systems. This is an example of why we describe the systems reform summarised in Table 3.1 as a narrow framing and one that provides an insufficient challenge to the way systems are configured. As such, it represents a continuation of a policy approach that has persistently failed to reduce spatial inequalities.

The nature of the political economy has influence over the systems logic (see Fig. 1.1) at different levels and is a key factor in the types of spatial inequalities that manifest. In Lansley's (2022) 200-year history of Britain's economic and social policy, he highlights the dominance of private ownership, where elites secure an excessive slice of the economic cake, reducing resources for wages, investment and innovation. Similarly, Christophers (2020: xxiv) defines rent as 'income derived from ownership, possession or control of scarce assets under condition of limited or no competition.' Christophers (2020) argues that the problem with the UK political economy, which produces low and stagnant growth, high levels of income and wealth inequality and low levels of innovation and investment, is not sectoral but relates to rentierism.

Rent is about *having* rather than *doing* and Christophers (2020) highlights seven types of rent that feature prominently in the UK economy: financial, natural-resource, intellectual property, platform, contract, infrastructure, and land. If we take contract rent as an example, Christophers (2020) focusses on outsourcing and argues that the UK has been at the forefront of public sector outsourcing since the 1970s, as part of an underlying rationale of anti-statism. Local government was one of the first areas in the firing line when Thatcher came to power, with the introduction of Compulsory Competitive Tendering (CCT) for services in 1980 (Christophers, 2020). More generally, Christophers (2020) argues that the ideological conviction that the state should own, and do, as little as possible has featured strongly over the past 40 years. This has clear implications for the capacity of the public sector to Level Up. As Lansley (2022, pp. 232-3) concludes, 'prioritising private over public wealth, and the erosion of the commonwealth base, have been among the most regressive and socially damaging state-driven trends of the last half century.' Lansley's (2022) argument for a system of 'asset redistribution' through national and local citizen-owned wealth funds alongside a more equalising tax and benefit system would provide a more solid foundation to Level Up given the implications for levelling the playing field. However, there is a good deal of competitive and possessive individualism within the UK social contract (Wistow, 2022) that has generally geared the political economy towards the interests of those in positions of wealth and power and been antagonistic towards intervening in the system in a redistributive manner, albeit with the exception of the post-war anomalist phase.

Capitals

In this section, we focus on the framework of 'six capitals' (human, financial, social, physical, intangible, and institutional) proposed in the White Paper (HM Government, 2022) that are concerned with capturing and evaluating the drivers of geographical disparities. These capitals are identified as individually significant but more importantly act in 'mutually reinforcing' ways to either create 'virtuous circles' or 'vicious and self-reinforcing cycles' (HM Government, 2022). As such, they play a central role in the government's conceptualisation of the 'left behind' problem and its plan to Level Up these places. However, there are two key respects the approach to capitals developed in the White Paper is flawed, relative to its wider objectives. Firstly, the understanding of these capitals is narrow and isolated from a full systems perspective and does not meaningfully embed the notion of accumulated capital and what this means for fair competition on a level playing field. Secondly, the track record and ideology of the Conservative Party is generally antagonistic towards the level of redistribution required to Level Up across these capitals.

In the forward to the LUWP (HM Government, 2022: viii) the former Prime Minister—Boris Johnson—stated 'the answer to it [left behind places] lies not in cutting down the tall poppies, or attempting to hobble the areas that are doing well.' This is entirely consistent with the contemporary political philosophy of the Conservative Party in which competitive and possessive individualism are comfortable bedfellows. However, this returns us to a key issue that the Levelling Up agenda is seeking to address but fails to fully understand—the cumulative impact of spatial inequalities in the UK and the implications of these for spatial disparities. This problem and the response to the problem are both longstanding, with Martin et al. (2021) highlighting that there has been a failure of UK governments since World War Two to grasp the importance of geographical inequality as a national problem. They (2021) argue that this is striking given the pioneering work of the Barlow Commission (1943) in recognising the interrelationship between the development of prosperous and poorer regions, including the dominance of London. The LUWP accepts that geographical differences emerge from structural transitions and that market forces, when left to their own devices, as has too often been the case in the UK, can lead to market failures that are socially unjust and economically uneven. Two types of market failure are identified in the White Paper (HM Government, 2022): those affecting left-behind places (self-reinforcing low growth, poor health and low well-being) and those afflicting well-performing places (the side-effects of success, including high house prices, long commutes and pollution). A focus on market failure alone is insufficient but HM Government (2022, p. 96) is clear that this does not extend to the prioritisation of redistribution strategies:

'By addressing place-based market failures, place-based strategies can grow the pie. They are about unleashing opportunity and boosting allocative efficiency, not redistribution between places per se. That is the essence of levelling up.'

This returns us to the quote from Johnson above, which represents a position that fails to integrate a highly significant factor in the UK's spatial disparities, i.e., the unusual influence of the core over the periphery in the UK's 'left behind' problem *and* the role of this in reproducing what is being levelled up to. We have already seen that Martin et al. (2021, p. 108) describe this as a 'spatially and systemically entrenched' problem. We might view it as the core moving at such a fast pace (albeit with internal dynamics and inequalities) that the rest (differentially) fail to keep pace with. In this respect, narrowing inequalities represents a dynamic and moving target with those performing less well (which contain much less favourable contextual and compositional characteristics) having to progress better, or at a faster rate, than those that are performing well (with favourable contextual and compositional characteristics including alignment with, and influence over, the political economy). Taking redistribution off the table reduces the levers available for any policy programme seeking to narrow inequalities.

From our discussions with the Directors of Regeneration, two from the 'Core City' and a 'London Borough' were somewhat aligned (albeit with strong qualifications and appreciation of spatial differentiation at different geographical scales) with the multiplier effects of agglomeration. Those from more peripheral locales were more critical, arguing that in industrial and economic policy the national narrative continued to be to "back the winners" and this needed: "Turning on its head. You need to back the losers more than the winners in Levelling Up." However, as the Director of Regeneration for a Midlands 'Medium Town' commented:

"It is a very brave politician at any level to say we will not in invest in London or the greater Southeast hotspot areas."

They continued:

"What you need to do is slowdown the overheating parts and allow the rest to catch up and overtime sort of come closer... there's an irony though in the whole White Paper...[which is] really actually quite good on analysis...But then all of the policy levers and stuff that it says the government will do are really marginal compared to the scale of the challenge that it sets out."

Bourdieu's (1986) notion of accumulated capital can be usefully applied to understand the significance of capital of different types accumulated over time for spatial disparities and inequalities. In this respect, the process of 'left behind' places has been long in the making and is deeply embedded, leading Martin et al. (2021, p. 24) to conclude that narrowing 'spatial inequalities, in both relative and absolute terms, will require such places to grow *faster* than more prosperous places, and for an extended period of time.' When we relate this to the Government's approach to understanding and conceptualising capitals it provides a further example of what we described in the previous section as a 'narrowly framed approach to complexity theory'. The LUWP draws heavily on a research paper published by the Bennett Institute (Coyle, 2019) for its conceptualisation of capitals outlined in Table 3.2. Coyle (2019) makes the case to move beyond a narrow statistical lens centred on GDP for policy making to a focus on a range of economic assets that people need to fulfil their economic potential. However, there is a strong case to widen the conceptual lens further and include sociological perspectives, which have been put to good use in academic debate when analysing issues such as social mobility (for instance: Friedman & Laurison, 2019; Payne, 2017). This turns attention to protective mechanisms that stop people from falling down. In this respect, we might also question what prevents places from declining, which in turn, returns us to Bourdieu's (1986, p. 241) conclusion that accumulated capital and history 'makes the games of society—not least, the economic game—something other than simple games of chance.' As we have already seen the problem is deeply embedded in

Table 3.2 The Levelling Up 'Capitals' (HM Government, 2022)

Capital	*Definition/understanding*	*Indicators/measures*
Human	Stock of knowledge, skills, competencies, health and other attributes.	Individual lifetime income; qualification level; employment support allowance; overweight and obesity prevalence; and smoking rates.
Financial	Financial activity contributing to spatial differences in productivity, jobs and living standards.	Share of equity and debt finance compared to GVA; distribution of equity; and location of investor-investee pairings.
Social	Social infrastructure, connectedness, active and engaged communities, and local satisfaction.	Community needs and satisfaction with local area.
Physical	Transport; housing and digital.	Travel time to work; ratio of house-price to residence-based earnings; capital stock per worker; gross fixed capital formation compared to London; and foreign direct investment per workforce job.
Intangible	Comprised of formal (e.g., IP and patents) and informal (e.g., embedded processes and practices) forms of investment.	Income from IP; patent activity and high-tech patents; STEM and high-tech employment; management practices score; and R&D expenditure.
Institutional	Degree of centralisation.	Percentage of adults who agree they can influence decisions affecting their local areas; social fabric score; and strength of LEP networks.

place, which can be interpreted as a result of differential accumulated capital in place that cuts across economic, social, cultural, organisational and infrastructural systems.

Table 3.2 summarises each of the capitals in the LUWP and the proxy measures that are included for each of these. The selection, definition and measures of the capitals are open to debate and critique, but they provide a framework for the development of a cross-sectoral approach and partially respond to some of the prior criticisms about a lack of clear goals and measures of success and failure, albeit within the parameters and limitations of wider government policy. However, the section on the 'interdependence between the capitals' (HM Government, 2022, pp. 89–95) focuses almost exclusively on push-pull factors associated with labour force migration. Undoubtedly, this is a significant issue, and the White Paper develops a view of capital that is dynamic, relational and interdependent. For example, connecting social, human, and physical capitals and how these contribute to attracting graduates; but it severely undervalues the implications of accumulated capital for the potential to Level Up. Jones (2019) argues that a lot is expected of skills in the global economy and that whilst evidence suggests supply-side initiatives can make a difference at a regional level, the governance and nature of state intervention at a regional level is often unable to plug skills and employment gaps, given the deeply economic nature of the problem.

Regarding developing human capital, the Director of Regeneration in the Northern 'other city' described how investment over the past 15 years has mainly focused on large capital investments but that these require skills that are not necessarily held within the city. They continued to argue that reskilling people requires revenue related activity and in the context of austerity local government had very little room for manoeuvre, especially given the UK Shared Prosperity Fund's funding for skills was not available until year three. More fundamentally, a lack of connectivity and stocks of brownfield land can reduce the viability of bids under funding criteria, which draws heavily on the Treasury Green Book and an approach that favours agglomeration through its cost-benefit criteria. For example, the Directors of Regeneration in the 'medium town' and 'large town' worked in post-industrial areas on the periphery of large conurbations but with limited connectivity and physical capital. They described the high costs and relatively low value for regenerating brownfield sites when compared with projects in neighbouring authorities with less deprivation. They also stressed that the capacity funding on offer through the Levelling Up Fund was: "not nearly enough to resource what is required to respond to these opportunities."

In not dealing with accumulated capital, the LUWP underestimates the depths of the roots of spatial inequalities and a somewhat relational view of capitals is undermined by this significant omission. As argued above, the approach to capitals could have been significantly enhanced through the inclusion of additional literature. For example, the UK 2070 Commission (2020) provides a historical analysis of spatial disparities and a high-level approach to Level Up out of the COVID-19 crisis. Marmot et al. (2020a, 2020b) created a contemporary evidence-base around health inequalities that has a good deal of overlap with the Levelling Up agenda given the significance of history, socio-economic inequalities, place, and post-industrialism and emphasises how austerity and COVID-19 exacerbated these issues. Finally, Martin et al.'s (2021) longitudinal study identifies what constitutes a 'left behind' place. Each of these studies tells us something about inequalities in time and space, suggesting these are a deeply entrenched feature of UK society. Therefore, they effectively conceptualise the Levelling Up problem as a much greater issue to respond to than the government does. In the final sub-section of this chapter, we turn to how the government has identified key measures of success in its 'missions', which provides a framework for determining whether a place has levelled up or not.

Missions and the Policy Programme

The LUWP (2022) sets out 12 missions (see Table 3.3) that act as objectives for an overall policy regime for the Government to deliver against. The approach (HM Government, 2022, p. 159) is described as the 'next steps in the policy programme', 'only part of the answer to levelling up', and 'significant, steppingstones on what must be a sustained journey of change.' As such, they are reasonably well qualified and the LUWP makes connections to broader government policy such as the 2021 Spending Review. The missions are divided into four objective areas: boost productivity, pay, jobs and living standards; spread opportunities and improve public services; restore a sense of community and local pride; and empower local leaders and communities. The missions do address, at least in part, criticisms of the agenda about a lack of clarity, measures and goals and represent some stretching targets across a diverse range of issues relating

Table 3.3 Levelling Up Missions and Objectives (HM Government, 2022)

Focus area	*Mission*
Boost productivity, pay, jobs and living standards by growing the private sector, especially in those places where they are lagging.	
Living standards	By 2030, pay, employment and productivity will have risen in every area of the UK, with each containing a globally competitive city, and the gap between the top performing and other areas closing.
Research & development (R&D)	By 2030, domestic public investment in R&D outside the Greater Southeast will increase by at least 40%, and over the Spending Review period by at least one third. This additional government funding will seek to leverage at least twice as much private sector investment over the long term to stimulate innovation and productivity growth.
Transport infrastructure	By 2030, local public transport connectivity across the country will be significantly closer to the standards of London, with improved services, simpler fares and integrated ticketing.
Digital connectivity	By 2030, the UK will have nationwide gigabit-capable broadband and 4G coverage, with 5G coverage for the majority of the population.
Spread opportunities and improve public services, especially in those places where they are weakest.	
Education	By 2030, the number of primary school children achieving the expected standard in reading, writing and maths will have significantly increased. In England, this will mean 90% of children will achieve the expected standard, and the percentage of children meeting the expected standard in the worst performing areas will have increased by over a third.
Skills	By 2030, the number of people successfully completing high-quality skills training will have significantly increased in every area of the UK. In England, this will lead to 200,000 more people successfully completing high-quality skills training annually, driven by 80,000 more people completing courses in the lowest skilled areas.
Health	By 2030, the gap in Healthy Life Expectancy (HLE) between local areas where it is highest and lowest will have narrowed, and by 2035 HLE will rise by five years.
Well-being	By 2030, well-being will have improved in every area of the UK, with the gap between top performing and other areas closing.

(*continued*)

Table 3.3 (continued)

Focus area	*Mission*
Restore a sense of community, local pride and belonging, especially in those places where they have been lost.	
Pride in place	By 2030, pride in place, such as people's satisfaction with their town centre and engagement in local culture and community, will have risen in every area of the UK, with the gap between top performing and other areas closing.
Housing	By 2030, renters will have a secure path to ownership with the number of first-time buyers increasing in all areas; and the government's ambition is for the number of non-decent rented homes to have fallen by 50%, with the biggest improvements in the lowest performing areas.
Crime	By 2030, homicide, serious violence and neighbourhood crime will have fallen, focused on the worst affected areas.
Empower local leaders and communities, especially in those places lacking local agency.	
Local leadership	By 2030, every part of England that wants one will have a devolution deal with powers at or approaching the highest level of devolution and a simplified, long-term funding settlement.

to spatial inequalities. However, the extent to which the missions shape policy direction, let alone represent a meaningful framework for performance management at the local level, is questionable. Our discussions with the Directors of Regeneration suggests that where the missions align with local priorities these are useful hooks to hang local policy around but where they are not, they are viewed as somewhat additional and discretionary, given constrained capacity regarding officer time and local government funding.

We do not have space to examine each of these missions, though there is more discussion in the next chapter on local leadership in relation to 'local democracy and devolution'. For now, we will focus on the health mission. The mission itself is a good one and will be further developed in the planned Health Disparities White Paper (which is expected to be published in 2022). HLE is highly significant to the quality and duration of life and provides a good indicator to both develop and assess health policy in terms of population health and health inequalities. Furthermore, in seeking to reduce the gap in HLE between local areas where it is highest and lowest a place-based focus on inequalities as a 'moving target' has been developed, comparing and linking the outcomes between the best performing and worst performing areas. As Marmot (2022) argues the alignment between the objectives and the relationship between the missions are good and if these were all to be achieved it is highly likely health inequalities would reduce and health equity would improve. However, despite identifying a valuable priority area, there are weaknesses in the ways that the LUWP (2022) frames the associated key issues.

Firstly, the mission is unclear about what thresholds will be used to determine areas with the highest and lowest HLE. Secondly, the 'case for action' (HM Government, 2022, pp. 200–202) is too narrowly focused on individual behaviours and demographics while the 'policy programme' (HM Government, 2022, pp. 202–206) concentrates on public health interventions, food and diet, and diagnostic backlogs. Each of these areas are necessary but not sufficient to reduce health inequalities, given they do not address key causal factors that produce the social and economic inequalities in which health inequalities are reproduced (see, for example: Harvey, 2021; Lynch, 2017; Scambler & Scambler, 2015; Schrecker, 2017; Wistow et al., 2015). The health mission shares many similarities with the other focus areas (for example, 'education', 'well-being', 'pride in place') in that key causal factors associated with the political economy and its relationship to societal and place-based inequalities are marginal to the debate about how to close the gap between better and worse performing areas.

Regarding the funding commitments to support the agenda, there is very little in the LUWP that could be described as additional funding from those outlined in the section of this chapter on the agenda prior to the White Paper.[3] In our discussions with Directors of Regeneration they all (albeit to varying degrees) thought the intent and aims of the Levelling Up strategy were admirable. However, they all (again to varying degrees but with more consistency) questioned both how realistic the aims of the agenda were and the validity of the missions' given timescales, resources and powers available through the strategy. Marmot (2022, p. 1) goes further in stating 'the problem is that this white paper reads as though it was not the product of a political party that has been in power in Britain for 30 of the past 43 years and is responsible for much of the damage.' In short, the trajectory of the neoliberal political economy, as (amongst other things) generally non-redistributive, competitive, and anti-state intervention in the socio-economic interests of the citizenry are key system

[3] Although in April 2022, the £2.6bn UK Shared Prosperity Fund was launched, which was designed to support the four high level objective areas outlined in Table 3.3.

features, in which those places higher up the social and economic distribution are not only protected from but more able to enjoy the benefits of a system geared towards competitive and possessive individualism. Wistow (2022) describes the political economy as both a protective and pathogenic mechanism—depending on where you sit in the social distribution—in relation to health inequalities. We might also consider that the political economy fulfils a similar function relative to place more generally.

Conclusion

This chapter explored the Levelling Up agenda and contextualised it within wider debates about political economy, spatial inequalities and regional and regeneration policy. Whilst offering some promise, the Levelling Up agenda is narrowly framed despite adopting a systems focus. In particular, the White Paper does not fully develop the implications of the nature of the political economy relative to the spatial disparities it (re)produces. While it was unlikely that a Conservative government would seek to radically change the nature of the political economy, the rhetoric and high-level policy goals emerging from the agenda suggest this is what is needed. On the one hand the agenda is implicitly redistributive but on the other it offers little in the way of systematic redistribution to enable 'left behind' places to catch-up with what is being levelled up to. Furthermore, the additional funding mechanisms provided through the agenda are likely to be insufficient given the scale of the 'left behind' problem and do not match the ambition of the agenda, as set out in the missions. Compounding this, the competitive-bidding funding model has been criticised for a tendency towards 'pork barrel' politics and an insufficient focus on deprivation. Given the argument above, the Levelling Up agenda and White Paper (HM Government, 2022) are unlikely to disrupt spatial trajectories sufficiently to prevent a continuation of high levels of spatial inequalities and, therefore, appears to represent a continuation in place-based policy failing to adequately address these. In the next chapter, we turn to the role of local government in local governance systems; in relation to local democracy and devolution and its role relative to Levelling Up, suggesting that the 'hollowing out' of local government under neoliberalism is not conducive to achieving the ambitions set out in the LUWP.

References

Argawal, S., & Phillips, D. (2020). *Catching up or falling behind? Geographical inequalities in the UK and how they have changed in recent years.* London Institute for Fiscal Studies.

Atkinson, R. (2020). *Alpha City.* Verso.

Bourdieu, P. (1986). The forms of capital. In J. Richardson (Ed.), *Handbook of theory and research for the sociology of education* (pp. 241–258). Westport, CT.

Carr-West, J., & Sillett, J. (2021). *On the level: Six principles to underpin the levelling up white paper.* LGiU.

Christophers, B. (2020). *Rentier capitalism: Who owns the economy, and who pays for it?* Verso.

Connolly, J., Pyper, R., & van der Zwet, D. (2021). Governing 'levelling up' in the UK: Challenges and prospects. *Contemporary Social Science., 16*(5), 523–537.

Copeland, P., & Diamond, P. (2022). From EU structural funds to levelling up: Empty signifiers, ungrounded statism and English regional policy. *Local Economy., 37*(1–2), 34–49.

Coyle, D. (2019). *Measuring wealth, delivering prosperity.* Bennett Institute for Public Policy.

Cross, K., Evans, J., MacLeavy, J., & Manley, D. (2022). Analysing the socio-economic impacts of COVID-19: A new regional geography or pandemic enhanced inequalities? *Regional Studies, Regional Science, 9*(1), 461–485.

Financial Times. (2021). *Levelling Up Fund bias in favour of Tory seats 'pretty blatant'.* Accessed on 05.06.2022. Available at: https://www.ft.com/content/d485da2a-5778-45ae-9fa8-ca024bc8bbcf

Friedman, S., & Laurison, D. (2019). *The class ceiling: Why it pays to be privileged.* Policy Press.

Good Law Project. (2021). *Pork Barrel Politics.* Accessed on 07.06.2022. Available at: https://goodlawproject.org/news/pork-barrel-politics/

Harvey, M. (2021). The political economy of health: Revisiting its Marxian origins to address 21st-century health inequalities. *American Journal of Public Health, 111*(2), 293–300.

HM Government. (2021). *Levelling up fund: Prospectus.* Her Majesty's Stationery Office.

HM Government. (2022). *Levelling up: Levelling up the United Kingdom.* Her Majesty's Stationery Office.

House of Commons Business, Energy and Industrial Strategy Committee. (2021). *Third report—Post-pandemic economic growth: Levelling up.* House of Commons.

Jennings, W., McKay, L., & Stoker, G. (2021). The politics of levelling up. *The Political Quarterly, 92*, 302–311.

Jones, M. (2019). *Cities and regions in crisis: The political economy of sub-National Economic Development.* Edward Elgar Publishing.

Lansley, S. (2022). *The richer the poorer: How Britain enriched the few and failed the poor. A 200-year history*. Policy Press.
Leyshon, A. (2021). Economic geography I: Uneven development, 'left behind places' and 'levelling up' in a time of crisis. *Progress in Human Geography, 45*(6), 1678–1691.
Liddle, J., Shutt, J., & Addidle, G. (2022). Editorial: Levelling up the United Kingdom? A useful mantra but too little substance or delivery? *Local Economy, 37*(1–2), 3–12.
Lynch, J. (2017). Reframing inequality? The health inequalities turn as a dangerous frame shift. *Journal of Public Health, 39*(4), 653–660.
Marmot, M. (2022). The government's levelling up plan: A missed opportunity. *British Medical Journal, 1-2*, 1.
Marmot, M., Allen, J., Boyce, T., Goldblatt, P., & Morrison, J. (2020a). *Health equity in England: The Marmot review 10 years on*. Institute of Health Equity.
Marmot, M., Allen, J., Boyce, T., Goldblatt, P., & Morrison, J. (2020b). *Build back fairer: The COVID-19 Marmot review*. Institute of Health Equity.
Martin, R. (2015). Rebalancing the spatial economy: The challenge for regional theory. *Territory, Politics, Governance, 3*(3), 235–272.
Martin, R., Gardiner, B., Pike, A., Sunley, P., & Tyler, P. (2021). *Levelling up left behind places: The scale and nature of the economic and policy challenge*. Routledge.
Payne, G. (2017). *The new social mobility: How politicians got it wrong*. Policy Press.
Scambler, G., & Scambler, S. (2015). Theorizing health inequalities: The untapped potential of dialectical critical realism. *Social Theory and Health, 13*(3–4), 340–354.
Schrecker, T. (2017). Was Mackenbach right? Towards a practical political science of redistribution and health inequalities. *Health and Place, 46*, 293–299.
Shearer, E., Shepley, P., & Soter, T. (2021). *Levelling up: Five questions about what the government means by the phrase*. Institute for Government.
Telford, L. (2022). *English nationalism and its ghost towns*. Routledge.
Tomaney, J., & Pike, A. (2020). Levelling Up? *The Political Quarterly, 91*(1), 43–48.
Travers, T., Sims, S., & Bosetti, N. (2016). *Housing and inequality in London*. Centre for London.
UK2070 Commission. (2020). *Go big. Go local: The UK2070 report on a new deal for levelling up the United Kingdom*. UK2070.
Wistow, J. (2022). *Social policy, political economy and the social contract*. Policy Press.
Wistow, J., Blackman, T., Byrne, D., & Wistow, G. (2015). *Studying health inequalities: An applied approach*. Policy Press.

CHAPTER 4

Local Government, Governance and Levelling Up

Abstract This chapter focuses on the role of local government relative to the Levelling Up agenda and in tackling spatial inequalities more generally. We deploy further data from the Directors of Regeneration regarding local governance and Levelling Up. The chapter considers the trajectory of local government given changes to its role through processes such as the contracting out of services and following a period of sustained cuts in funding during the austerity era. We also consider debates about the function of local government in relation to local democracy and devolution, concluding the chapter by returning to the implications of the Levelling Up agenda for local government.

Keywords Local government • Governance • Levelling Up

Introduction

As is hopefully clear to the reader, the Levelling Up agenda seeks to influence governance systems at a variety of levels of spatial aggregation including regional, sub-regional, local and neighbourhood. In this chapter, we are concerned with each of these but from the 'entry point' of local government. The reason for this is that despite many years of decline in terms of its functions and resourcing local government is still the key state

L. Telford, J. Wistow, *Levelling Up the UK Economy*,
https://doi.org/10.1007/978-3-031-17507-7_4

institution of relevance to Levelling Up's aims and, perhaps more importantly, is the principal site of local democracy in the UK. In fact, this democratic function provides not only accountability and legitimacy for local authorities to act on behalf of their populations but also the reach and knowledge that other institutions lack. A key argument that will be developed is that these organisations have been marginalised and underutilised with the COVID-19 pandemic providing a prime example, albeit one in which local government played a highly significant role largely despite, rather than because of, central government support and engagement, in the highly centralised system of UK government and policy.

The chapter begins by highlighting the role of local government in local governance systems. The decline of local government over the past 40 years will be explored in relation to declining function and powers; funding constraints; the move to commissioning and contracting out of services; and changes in regional and sub-regional government (see for example: Barnett et al., 2021; Hambleton, 2020). These contribute to the growing significance of local governance systems in which the balance has shifted to local authorities commissioning and co-ordinating services rather than directly providing them, with implications for the role and capacity of local government to fulfil its democratic function. At times, issues that have been discussed in previous chapters such as the welfare state, deindustrialisation and austerity are briefly reintroduced particularly in relation to local government and their lack of capacity/resources to address the consequences. The chapter also considers the trends identified within local governance systems alongside debates about devolution and local democracy. It concludes by considering the role and prospects of local government given the implications of the Levelling Up agenda. In doing so, the first part of the chapter provides context about local governance systems, while the second part integrates this with data from the Directors of Regeneration.

Local Government and Governance Systems: Trajectories of Decline and Fragmentation

As Chap. 2 documented, the post-war period was a time of rising living standards, relative prosperity and a sizable increase in working class bargaining power. The welfare state was a core component of this, with local government playing a significant role in its development and delivery. Newman (2014) highlights that councils had a substantial remit in planning, place-shaping and the delivery of education, housing and social services, which led Jones (2019) to describe local government as the formal partner of central government in the Keynesian welfare state. Over the past 40 years there has been a considerable shift in power from local to central government (Leach et al., 2018), which Jessop (2015) characterises as a form of path dependency resulting from the neoliberal redrawing of boundaries between state and non-state actors. Peck (2003) described these tendencies as 'roll out' and 'roll back' neoliberalism, in which the state's apparatus are used to create space for private capital through reductions in, and the privatisation of, public services (also see: Mitchell & Fazi, 2017; Streeck, 2016). Jones (2019) argues that the New Right was successful in constructing local economic failure as a result of state involvement and confining the role of the state to maintaining the conditions for the market to function. Therefore, Hill and Hupe (2014) conclude that neoliberal government policy and a strong preference to contract out organisational activities has reduced its own service delivery capacity. Through an increasingly centralised state there has been (somewhat paradoxically) growing fragmentation and complexity in the local state, resulting from an increased plurality of self-organising actors in local systems, which is a feature of and contributor to the system logic outlined in Fig. 1.1. For example, Jones (2019) argues that central government has retained control through normalising neoliberalism by widening local authority spatial coalitions and drawing local government into an economic agenda centred on the primacy of the market, deregulation and privatisation.

In effect, there has been a move from government to governance (Rhodes, 1997), in which local authorities face growing difficulties in maintaining strategic oversight in key services, given their responsibilities have been 'hollowed out' and they are one provider among many (Barnett et al. (2021). Hambleton (2020, p. 56)) describes government and governance respectively as:

> the formal institutions of the state. Government makes decisions within specific administrative and legal frameworks, and uses public resources in a financially accountable way.

> involving government *plus* the looser processes of influencing and negotiating with a range of public and private agencies to achieve desired outcomes. A governance perspective recognises the importance of collaboration between the public; private and non-profit sectors to achieve mutual goals.

We have already seen that local government has been at the forefront of the transition to governance. The introduction of CCT in the 1980s was a significant mechanism in changing the role of local government to commissioners of services as opposed to being a provider of these. More recently, the coalition government's *Open public services* White Paper (HM Government, 2011) made it clear that the default position was the state should commission services from a diverse range of providers in the private and voluntary sectors. The 2014 Care Act similarly set out that local authorities are responsible for commissioning care (NAO, 2021), as opposed to directly delivering this. Adult social care is the area in which contracting out of services is perhaps most developed. There are now 14,800 registered providers across 25,800 locations in England (NAO, 2021). As Wistow (2022, p. 65) argues, 'the growing complexity in governing and governance stems from and is intensified by, the increased diversity of partners involved in service provision in neoliberal governance systems, generating new non-linear dynamics and fragmentation resulting from increased plurality of self-organising actors.' This is part of the macro and meso context in which adult social care will make its contributions to the 'health' and 'well-being' missions of the Levelling Up agenda. As Jessop (2016) argues, the complex set of social relations in governance systems extends beyond the state system and its capabilities with implications for the power of state managers to exercise power over these.

To the complexity and fragmentation identified above, a further feature of governance systems is the reduced capacity of local government as an actor. Contracting out of services is itself a reduction in both service delivery capacity and functional capacity. In this respect, Christophers (2018) highlights that an early act from the 1979 Thatcher government was to terminate the Redundant Lands and Accommodation Act, thereby removing local government's first option to develop this land before the private sector had a chance to purchase it. There are clear implications for a reduction in the place-shaping role of local government. A further key and ideologically and politically motivated area (see: Broughton, 2019; Minton, 2019; Slater, 2018) in which local government has lost capacity within spatial policy is in social housing. Broughton (2019) argues that council housing was seen as providing the wrong kind of accountability (i.e., to local government and tenant associations, instead of to the market) and as limiting choice and personal consumption. The 1980 Housing Act massively accelerated council house tenants' 'right to buy' their homes, with Minton (2019) arguing this was the most important privatisation of the neoliberal era. 1.8 million council homes had been sold by 1997 and council house building declined dramatically in that period from 79,160 homes in 1978–79 to 400 in 1996–97 (Broughton, 2019).

'Right to buy' represents a significant loss of capacity for local government in its role in providing social housing and has contributed to a concentration of social problems in low value privately rented accommodation. For example, Lloyd et al. (2021) highlight how low-cost housing can lead to competition for scarce resources in specific wards and neighbourhoods within a local authority area in the Northeast with high levels of multiple deprivation. The Director of Regeneration for the 'Northern Core City' similarly described how low-cost poor-quality housing can concentrate in areas of "real poverty" within the city, with overcrowding, safeguarding issues for young people (including potential trafficking), poor public health sanitary conditions, and the designation of these as places for the resettlement of refugees. The LUWP's missions on 'pride in place' and 'housing' (see Table 3.3) includes a focus on poor quality housing and a target to reduce the number of non-decent homes by 50% by 2030. However, beyond plans to explore a new national landlord register and a Social Housing Regulation Bill which will enable residents to hold landlords to account, there is little in the LUWP that suggests local government will be enabled to make transformational change to concentrations of deprivation centred around low-cost and poor-quality housing. As Beel et al. (2021) highlight, housing is an area in which market logic has been incorporated into policies promoting commodification, privatisation and financialisaton. These all impact on the ability of the local state to respond to citizens' needs, given the primacy of the market.

Unsurprisingly, the 2010 to 2020 period of austerity also played a very significant role in reducing the capacity of local government as an actor in local governance systems. Alongside the other cuts outlined in Chap. 2, there was a 29% reduction in local government real-term spending power between 2010–11 and 2019–20 (NAO, 2021). Phillips and Simpson (2018) highlight the highly regressive nature of these cuts in a similar period (2009–10 to 2017–18) as funding per person was reduced by 35% for those living in local authorities in the most deprived quintile, whereas for those in the least deprived quintile it only reduced by 15%. At a higher level of aggregation, Marmot (2022) highlights the post 2010 cuts to local government, amounting to £413 per person in the North, which he contrasts with an allocation of £32 per person in the North in 2021 under Levelling Up. Given the depth of the problem of 'left behind' places and the sustained spatial trajectory of austerity, it is questionable whether the Levelling Up funding mechanisms (discussed in Chap. 3 and again later in this chapter) will change this trajectory, particularly as the allocation criteria for these competitive-funding schemes do not fully account for deprivation.

More generally, local government functions within a wider sub-national governance system that experienced a great deal of churn over the past forty years with changing regional and sub-regional institutional arrangements (Jones, 2019; Martin et al., 2021; Pike et al., 2018; Wistow, 2022). Copeland and Diamond (2022) situate this within Dunleavy's (1989) notion of ungrounded statism because government exercises financial control from the centre rather than driving policy and innovation from the regional and local level. Copeland and Diamond (2022) view English governance as both highly centralised, fragmented and differentiated with a weakness and lack of capacity for public institutions at the sub-regional level. The LUWP (HM Government, 2022, p. 133) states that:

> In England, local governance is split across county councils, district councils, unitary authorities and London borough councils and, at the sub-regional level, mayoral and non-mayoral combined authorities and the Greater London Authority (GLA). Local Enterprise Partnerships and Pan-Regional Partnerships (PRPs) also have some responsibility for economic strategy.

Connolly et al. (2021) argue that such fuzzy governance arrangements are convenient for those seeking to avoid being held to account. Similarly, Beel et al. (2021) highlighted concerns about accountability and the marginalisation of local civil society through the development of Local Enterprise Partnerships (LEPs). Following the abolition of the nine Regional Development Agencies in 2010, 38 LEPs have been introduced at the sub regional level. LEPs were created to enact a rebalancing towards and privileging of the private sector through integrating business leaders and their knowledge and networks into this level of governance (Newman & Gilbert, 2022). As Newman and Gilbert (2022) argue, there are signs these will be subsumed under new city and county deals but there are important lessons to learn. In particular, Newman and Gilbert's (2022) research highlights four lessons for future policy design:

- An assumed cleavage between the business and civic context was incorrect given the networks councillors and public sector employees have with the private sector;
- LEPs imported partial perspectives about the private sector's knowledge about local economies that was interpreted by local leaders as full insights;
- The connectedness to business networks led to concerns about conflicts of interest and opaque decision-making; and
- The presence of appointed elected officials (as opposed to being led by an elected body) on LEPs did not address a significant deficit in democratic accountability.

In summary, they (2022, p. 16) claim 'it is clear that a business-led model of subnational governance tends to deviate from representativeness, transparency and accountability, creating challenges for institutional design.' Similarly, Copeland and Diamond (2022) highlight that they are not political or administrative institutions, which has contributed to a governance deficit in English sub-national economic development. In the next section, we consider issues relating to local democracy and devolution and the response of the Levelling Up agenda to ongoing challenges to institutional design at the local level. For now, it is clear from the LUWP (HM Government, 2022, p. 235) that private sector-led partnerships will have a role not only to 'coordinate action across the private sector, but in partnership with local and central government, and local education and research institutions.'

Local Democracy and Devolution

As stated, local government has a significant function as the primary arm of local democracy, providing a degree of representation and accountability for the public at the local level. The devolution of funding and powers to local government offers the potential to resolve some of the contradictions and fragmentation in local governance systems through strengthening the role of local government and by association local democracy, given the direct line of democratic accountability provided (see, for example: Hambleton, 2020; Tomaney, 2016; UK2070 Commission, 2020). However, it is important to recognise that local elections have been described as 'second-order' elections because voters see them as less important than national elections (Heath et al., 2003). Furthermore, the period in which local government has declined and local governance has emerged has not only complicated the role of local government but also undermined its legitimacy through a loss of capacity to respond to local needs and preferences. The UK2070 Commission (2020) links this to the highly centralised nature of decision-making in the UK that relies upon eroded local capacity to take action and the confusion about what is a national decision and what are local decisions. However, as Skelcher (2003) emphasised local democratic politics still revolves around local councils, despite agencies of local public policy becoming more highly differentiated and multi-layered.

Hambleton (2020) argues that globally, elected local leaders are operating in a context of place-less power, in which decisions made by MNCs, for example, are unconcerned about the impacts on local communities, thereby placing a limit on what can be achieved. Such a dynamic was highlighted in Chap. 2, with deindustrialisation in the UK leading to branch plants moving to low wage economies impacting detrimentally on R&C in Teesside as well as other 'left behind' localities (MacLeavy & Jones, 2021; Rodriguez-Pose, 2018; Sandbu, 2020; Tomaney et al., 2021). This, in turn, has co-evolved with wider processes influencing democratic participation. In the 1970s overall turnout in local elections generally exceeded 40% of the electorate; but by the 2010s this had declined to around 30% (Uberoi, 2021). Nevertheless, despite the constraints and limitations facing local democracy across complex multi-level systems it remains an important and underutilised avenue to understand and respond to place-based needs and demands. For example, Hambleton (2020, pp. 12–14) identifies three areas in which place matters:

- It forms an important part of our human identity through our physical relationships with family, friends, neighbours and colleagues.
- It provides a focus for differential needs that government responsiveness can be tuned to.
- It provides the spatial unit for the exercise of democracy.

These are all areas in which local government can fulfil an important function. We are, however, a long way short of place-based policy in the UK given the centralised nature of the system and the lack of resources and functional capabilities for local government as the arm of local democracy. Martin et al. (2021) defines place-based policy as bottom-up locally-led policy, tailoring public goods and services to local and regional problems and potentialities. The COVID-19 pandemic provides an excellent example of the relationship between central and local government and the unwillingness of the former to trust and resource the latter despite the development of the nascent Levelling Up agenda. For example, the NAO (2020) highlighted the experience of local authorities in track-and trace and their statutory duty to control local disease outbreaks. However, the UK did not follow international trends to involve local government in these services, instead preferring to contract it out to the private sector around a call centre model that, at times, was barely used (NAO, 2020). This can be interpreted as an example of the rentier and neoliberal system logic at the macro level cutting across hierarchical systems. Wistow (2022) argued that an opportunity to restore much needed capacity to local government was missed due to a narrow ideological preference for a small state, even during widespread spending of public money. To do so at a time when Levelling Up was a significant component of the government's policy-rhetoric should call into question several aspirations of this agenda, particularly around empowering local leaders and communities. Despite the pandemic demonstrating a capacity for action, if there is a common purpose, with public and private sectors coming together (UK2070 Commission, 2020), the UK's approach to track and trace represents a missed opportunity to invest in and support local government. This was short-sighted because enhancing the capacity and reach of the local state in this respect could have increased the knowledge about, and connections to, local populations during a crisis. These are vital assets relative to the challenges of deprivation, hard-to-reach and 'left behind' populations and Levelling Up. Indeed, the White Paper (HM Government, 2022, p. 233) subsequently stated that:

> achieving the UK Government's levelling up missions will rely on local leaders being empowered to develop local solutions to local problems. Many places have large gaps in physical, human and social capital. But the precise challenges they have to deal with are unique to their locality. Strong local institutions and leadership are needed to tackle these local problems effectively on a targeted and sustained basis.

As we will see in the next chapter, some of these challenges include employment insecurity, the degeneration of community life, crime and anti-social behaviour. Nevertheless, as Broadhurst and Gray (2022, p. 11) demonstrate despite the Levelling Up rhetoric, 'each government department continued to operate in silos without a unified definition of place…favouring city policy networks to gather feedback on the impact of the pandemic at the local level.' Carr-West and Sillett (2021) argued that following the pandemic there must be a new settlement for places linking the Levelling Up ambitions to the decentralisation of power. The LUWP (HM Government, 2022, p. xix) is the focal point of that settlement and has some fairly strong rhetoric in this respect:

> central government decision-making will be fundamentally reoriented to align policies with the levelling up agenda and hardwire spatial considerations across Whitehall. This will require greater transparency around the geographic allocation of funding and simplification of local growth funding. It will mean running levelling up through central government decision-making as a golden thread for which departments are held accountable. And it will mean extra resources being deployed to local areas, including moving 22,000 civil servants out of London by 2030.

The movement of 22,000 civil servants across 15 government departments to places like Wolverhampton, Darlington, Tyneside, East Kilbride, Cardiff, Belfast, Sheffield, and Bristol is a positive step given the direct and indirect impact of relatively highly skilled and renumerated jobs on local employment. It also represents a broadening of the lens and focus of the civil service given the significance of place to experiential knowledge. However, the extent to which this truly devolves power away from Whitehall remains to be seen given civil servants will continue to report to a centralised and hierarchical bureaucratic system tied to the Westminster model of government. In short, the movement of these jobs is welcome, but it does not represent a shift to locally led policy and instead diversifies the physical location of employment at the centre. Indeed, the 'hardwiring' of spatial considerations was met with some scepticism by the Director of Regeneration for the 'Northern Other City'. They stated:

> I keep seeing the dead hand of the Treasury all over it [funding allocation] and the civil service is having a dampening effect not just over local government but central government too.

Similarly, the Director of Regeneration in the 'Northern Core City' expressed their disappointment that there was very few new policy tools and instruments, arguing they would have to rely on existing tools. Given the significance of the macro-context for micro-level interactions, the need for new devolved powers to Level Up is clear. Indeed, Hambleton (2020) highlights how local leaders have a major advantage regarding being able to tap into local knowledge and understanding but that they are ill-equipped in the face of place-less power. By place-less power Hambleton (2020) is referring to the forces of globalisation and the exercise of power by decision makers including from MNCs, who are unconcerned about their impact on communities living in particular places as will be further discussed in the next chapter.

The Cities and Local Government Act 2016 set in train a move towards integrated governance through mayoral combined authorities and a focus on deal-based devolution through growth deals (Beel et al., 2021). The LUWP (HM Government, 2022, pp. 233–244) subsequently set out the government's plans to 'empower local leaders and communities' with warm rhetoric around empowering 'local leaders to develop local solutions to local problems'. The White Paper (HM Government, 2022) outlines plans to extend devolution to 11 counties through new devolution deals; deepen devolution through new 'trailblazer deals' for Greater Manchester and the West Midlands and work with other Mayoral Combined Authorities and the Greater London Assembly to streamline funding and bid for new powers; and simplify devolution through a new devolution framework to establish a new form of combined authority model to provide a single accountable institution for a functional economic area. The preferred option is for a directly elected leader of a combined authority but there is scope for different levels of devolution with functions relating to which model is pursued (HM Government, 2022). These levels are briefly summarised below:

- Level 1—local authorities working across a 'functional economic area' (FEA)—i.e., at the sub-regional level that the LEPs function at—and have the opportunity to host functions and pool services at a strategic level and to deliver action on climate change and Net Zero.
- Level 2—a single institution or County Council without a directly elected mayor covering an FEA or County Council. Functions include those at level 1 plus: transport planning, adult education functions and budget, the ability to introduce bus franchising, Home England compulsory purchase powers, and clearly defined roles in local resilience.
- Level 3—a single institution or County Council with a directly elected mayor covering an FEA or County Council. Functions

> include all of those at levels 1 and 2 plus: being prioritised for new rail partnerships, long-term investment fund, role in designing and delivering contracted employment programmes, ability to establish Mayoral Development Corporations, devolution of locally-led brownfield funding, strategic partnerships with Homes England over the Affordable Housing Programme, Mayoral control of Police and Crime Commissioner functions, a new duty for improving the public's health, and the ability to introduce mayoral precepting on council tax and a supplement on business rates (HM Government, 2022).

Our interviews highlighted how the combined authority route may not be attractive to all local authorities given historical political relations across neighbouring councils and the significance of electoral cycles for undermining long-term planning. The Director of Regeneration in the 'Northern Other City' described the LUWP's approach to devolution as, "devolution of power upwards [to a prospective combined authority], quite a stick to hit local politicians with—if you want more money, do as I say." They also cited uncertainty about what devolution-deals would lead to and how this made it a very hard sell to elected members, since it would mean surrendering some of their profile and powers to a combined authority. Therefore, the LUWP has been sensible in allowing for different paces of devolution. Even so the model was viewed as prescriptive by some of the Directors of Regeneration. They stated that despite the different levels of devolution the White Paper "rewarded the preferred model" and that this depended on how "clubbable" a local authority and its elected members were.

Those local authorities in our study that were members of combined authorities identified benefits of this through devolved powers and enhanced opportunities to access to funding. However, they also described issues with the allocation of funding across combined authorities. For example, the Director of Regeneration in the 'Midlands Medium Town' stated that when funding had been granted:

> The theory and intent behind the model disappeared and faded away and was suddenly forgotten about because everybody just wanted their pet project.

Similarly, in the 'Northern Large Town' the Director of Regeneration spoke about being able to secure the maximum allocation from the UK Shared Prosperity Fund for the city region through an investment plan based on calculating all the component district needs. However, when the funding was released central government said it was for the city region to determine how to allocate the money across its constituent parts. They stated that:

> It is devolved to the city region but not necessarily to where the need is…the system doesn't seem fair in terms of areas of need. It has been so frustrating that you almost lose the fight.

In summary, these Directors of Regeneration felt there was a good deal of ambiguity in the Levelling Up agenda around the internal working of devolution-deals and that there was a lack of guidance and engagement from central government about how to respond to internal dynamics and tensions around resource allocation within a combined authority. Once again, the significance of deprivation for these 'left behind' locales was raised in relation to within city-region inequalities and the implications to Level Up. However, deprivation was considered to have little weight in terms of allocating funding within a combined authority once it had been secured. The high-level mission around devolution (see Table 3.3) is explicit about the deal-based approach to devolution and that by 2030 all areas that want one will have one. As Beel et al. (2021) highlight deal-making public policy requires negotiation. Our findings suggest there is a good deal of fatigue around this kind of approach given constrained capacity, the impacts of years of austerity, and contradictions between Levelling Up's objectives and the dynamics between local partners within a competitive funding model.

Local Government and Levelling Up

In this section, we consider the implications of the Levelling Up agenda for the role of local government in tackling spatial disparities. In general terms, local government has been provided with a very bad deal from central government in 'left behind' places, given the relative lack of support to address post-industrial transitions (Byrne & Ruane, 2017; Jones, 2019) and the heightened levels of social and economic need resulting from this (Beatty & Fothergill, 2016). Local government is a key actor at the micro level and one that functions within a heterogenous welfare state in which there is variable capacity and demand at different levels (Warren & Wistow, 2017). Responding to heightened need resulting from the long-shadow of industrial destruction has fallen on local authorities with the most regressive cuts to funding under austerity. Nevertheless, Beel et al. (2021, p. 76) argue that 'the significance of local authorities as "anchor institutions" in terms of mitigating the impact of austerity and as a source of contestation in relation to negotiating and opposing austerity cannot be overstated.' The former is our primary concern here, alongside the role and capacity of local government to Level Up, though it is worth emphasising that these are related. For example, Streeck (2016) claimed privatising formerly public services reduces the material base for the legitimacy of states. In the previous section we saw that local government had been particularly susceptible to this change. Consequently, the potential for local government to act as a source of contestation has been reduced through its diminished capacity to respond to local need and the implications of this for its democratic legitimacy.

The role of local government relative to Levelling Up and the 'left behind' problem was nicely summarised by the Director of Regeneration for the 'Midlands Medium Town':

> You know the fascination in the trade press and the press generally is the Medici stuff, and, you know, taking the mick out of it. But actually, if you look at Germany and reunification and the effort to get, Eastern Germany, up to the level of Western Germany it took billions and billions a year, to make that happen and the trust that the German national government have in the local authority system. The ingredients you need are long term sustained money to places that need it, with trust in local actors to make the right local decisions, versus our heavily centralised state, with the competitive bidding beauty parade.

In the previous chapter, the funding model was discussed in relation to deprivation not featuring strongly in the allocation formula. We also briefly considered the nature of the funding streams, which are heavily skewed towards capital investment in physical symbols of Levelling Up. Indeed, The Director of Regeneration in the 'Northern Large Town' stated that:

> Those projects will be highly visible in the town centre and recognised through the Levelling Up branding, but will it really level up the issues? The answer is no.

In common with most of the Directors of Regeneration, they claimed additional revenue funding would be far better aligned with the complex nature of spatial disparities within and between local authority areas. This is close to the vision of the UK2070 Commission who, as we have seen, argued for £375 billion investment over 25 years to level up and crucially to make this part of a full devolution settlement to empower local leaders to take action. In the field of health inequalities, Marmot et al. (2020) developed the notion of 'proportionate universalism', which readily translates to the field of spatial inequalities. Their idea is to fund public services and invest in places relative to social need and health inequalities. Crisp et al. (2019) made a similar proposal in their research around local and regional development and suggested proportionate allocation of spend per head relative to the prosperity of places, recommending a £250 billion fund over ten years. These examples could empower local government to invest in the localised nature of spatial disparities through enhancing their revenue. As the UK2070 Commission (2020) argues, this has the potential to renew local democracy through strengthening the capacity of local government.

Increased revenue for local government to invest in place could aide local actors to shape their local economies in the face of global and national trajectories. However, currently the scope of local government to Level Up within the neoliberal and rentier political economy was also questioned by the Director of Regeneration for the 'Northern Other city'. They stated:

> we're providing an economic platform for people to be successful on and very little is generally felt in the local economy.

They described a polarisation between wealth creation and wealth experienced in the city, in which a number of Public Limited Companies were major employers but none of them reported their profits in the local authority area and many of the skilled employees lived outside the city in neighbouring rural areas and paid their council tax there. The Director of Regeneration argued that this compromised the multiplier effect in the city. It is also an example of the macro and meso system logic, highlighted in Fig. 1.1 and discussed more fully in Chap. 1, limiting the potential for local government to act at the micro level relative to the objectives of Levelling Up. The Levelling Up agenda (including the approach to systems reform summarised in Table 3.2) does not challenge this system logic, thus representing a continuation of what Copeland and Diamond (2022, p. 9) describe as a 'reliance on market forces to correct regional disparities' with 'projects scattered across the English regions while a strategy to coordinate them remains absent'.

All Directors of Regeneration taking part in our scoping study reported a lack of, or unequal, connectivity to the economy through history (e.g., labour market), physical infrastructure, skills and population dynamics. This contributed to a concentration of needs and social problems in smaller geographic areas that have to catch up. Whilst connectivity is key (and is a strong focus of the White Paper) it was not considered to be something that was within the control of local government. For example, the Director of Regeneration in the 'Northern Core City' reported that central government was more focused on discussions about outputs from town centre projects as opposed to strategic issues such as sub-regional connectivity in discussion with local government. They argued that they were not having a "single integrated conversation about place and sustainable growth". Similarly, the Director of Regeneration in the 'Northern Large Town' argued that the different pots of money available through Levelling Up are not joined-up, which undermines the capacity of local government to develop the type of systems reform the government is encouraging in the LUWP. Transport, in particular, was an area of concern raised by all the Directors of Regeneration. Their local authorities had varying degrees of competitive disadvantage through their connectivity to the wider transport network. In the case of the 'Northern Core City' this related to relatively poor connections within the national network, in particular, and to other parts of the region and intra-regional connections across the North of England. For the Director of Regeneration in the London Borough the relatively poor position of their local authority in the overground and underground network meant that it was much less able to access the benefits of economic development within the capital than Boroughs with better connectivity in the city-region. The Midland 'Medium Town' and Northern 'Large Town' were both peripheral within national and city-region networks. Both Directors of Regeneration argued that the allocation of transport funding was skewed particularly towards agglomerations and London. In this respect, Crisp et al. (2019) demonstrated how criteria for transport spending has historically favoured time savings, which is geared to reducing congestion for high paid workers rather than unlocking new opportunities for development.

In summary, there was a sense across our scoping study that local government was not being backed sufficiently in order to Level Up. There were a range of concerns from: a lack of new tools and resources provided to local government; insufficient funding available relative to the scale of the problem; a reliance on capital as opposed to revenue in the new funding initiatives; an insufficient consideration of deprivation across the agenda; and limited and unequal capacity across local government as a sector to compete for time limited and intensive funding schemes. It is also important to note that the London Director of Regeneration reported a sense of disinvestment in London amongst their peers in the capital.

Local government remains a key actor in responding to the Levelling Up 'left behind' places problem. The scale and depth of the problem (MacKinnon et al., 2022; Martin et al., 2021) occurs at the micro level; but as highlighted it is important to not view this as something that can be reduced to the micro level alone. As documented in Chap. 2, the depth and nature of the problem has co-evolved with the shift to a post-industrial and neoliberal political economy. Here we can turn to Titmuss (1968) who predates neoliberalism but nevertheless wrote about progress in societies and how this is not shared equally. He argued that socially caused 'diswelfares' are the price some people pay for other peoples' progress and considered the post-war welfare state to be a 'partial compensation' for these social costs and insecurities. In this light, it is difficult to interpret the Levelling Up agenda as anything other than a partial compensation for the disbenefits that have been accumulated in particular types of places (see Martin et al., 2021) over prolonged periods of time. It is hard to make the case that central government has equipped local government to adequately respond to this, let alone to fulfil its potential as a key place-shaping actor.

Conclusion

This chapter explored some of the implications of the Levelling Up agenda for local government *and* some of the implications of the changes to local government over the past forty years or so for its role relative to the Levelling Up agenda. From a complex systems perspective, the political economy has been highly significant over this period as the view that the state should intervene in the natural functioning of markets as little as possible has led to local government increasingly commissioning as opposed to delivering services. Over the past decade, austerity has hit local government particularly hard and further reduced its capacity as an actor in local governance systems. Approaches to devolution have occurred within this context. The LUWP provides some enhanced opportunities for local authorities willing to enter into combined authority structures. However, local dynamics between neighbouring areas and deal-based devolution are disincentives for some areas. The final section of the chapter considered the context in which local government functions relative to the Levelling Up agenda. It seems unlikely that local government has been provided with the funding or policy levers required to counteract the deeply embedded nature of the 'left behind' problem in the localities most in need of Levelling Up. In the next chapter, we turn to our findings from interviews with residents from 'left behind' R&C in Teesside. This provides an additional 'entry point' into the multi-scalar complex systems in which the Levelling Up agenda sits, while further elucidating the embedded challenges to Levelling Up.

References

Barnett, N., Giovannini, A., & Griggs, S. (2021). *Local government in England: 40 years of decline*. Unlock Democracy.

Beatty, C., & Fothergill, S. (2016). *The uneven impact of welfare reform: The financial losses to people and places*. Sheffield: Centre for Regional Economic and Social Research.

Beel, D., Jones, M., & Rees Jones, I. (2021). *City regions and devolution in the UK: The politics of representation*. Policy Press.

Broadhurst, K., & Gray, N. (2022). Understanding resilient places: Multi-level governance in times of crisis. *Local Economy, 37*(1–2), 84–103.

Broughton, J. (2019). *Municipal dreams: The rise and fall of council housing*. Verso.

Byrne, D., & Ruane, S. (2017). *Paying for the welfare state in the 21st century: Tax and spending in post-industrial societies*. Policy Press.

Carr-West, J., & Sillett, J. (2021). *On the level: Six principles to underpin the levelling up White Paper*. London: LGiU.

Christophers, B. (2018). *The new enclosure: The appropriation of public land in neoliberal Britain*. Verso.

Connolly, J., Pyper, R., & van der Zwet, A. (2021). Governing 'levelling up' in the UK: Challenges and prospects. *Contemporary Social Science, 16*(5), 523–537.

Copeland, P., & Diamond, P. (2022). From EU Structural Funds to Levelling Up: Empty signifiers, ungrounded statism and English regional policy. *Local Economy, 37*(1–2), 34–49.

Crisp, R., Ferrari, E., Fothergill, S., Gore, T., & Wells, P. (2019). *Strong economies, better places: Local and regional development for a Labour Government*. London: The Labour Party.

Dunleavy, P. (1989). The United Kingdom: Paradoxes of an ungrounded statism. Castles FG The Comparative History of Public Policy. Cambridge: Polity.

Hambleton, R. (2020). *Cities and communities beyond COVID-19: How local leadership can change our future for the better*. Bristol University Press.
Heath, A., McLean, I., Taylor, B., & Curtice, J. (2003). Between first and second order: A comparison of voting behaviour in European and local elections in Britain. *European Journal of Political Research, 35*(5), 389–414.
Hill, M., & Hupe, P. (2014). *Implementing public policy*. Sage.
HM Government. (2011). *Open public services: White paper*. The Stationery Office.
HM Government. (2022). *Levelling Up: Levelling Up the United Kingdom*. London: Her Majesty's Stationery Office.
Jessop, B. (2015). Crisis, crisis-management and state restructuring: What future for the state? *Policy & Politics, 43*(4), 475–492.
Jessop, B. (2016). *The state: Past, present, future*. Polity Press.
Jones, M. (2019). *Cities and regions in crisis: The political economy of sub-national economic development*. Edward Elgar Publishing.
Leach, S., Stewart, G., & Jones, G. (2018). *Centralisation, devolution and the future of local government in England*. Routledge.
Lloyd, A., Devanney, C., Wattis, L., & Bell, V. (2021). "Just tensions left, right and centre": Assessing the social impact of international migration on deindustrialized locale. *Ethnic and Racial Studies, 44*(15), 2794–2815.
MacKinnon, D., Kempton, L., O'Brien, P., Ormerod, E., Pike, A., & Tomaney, J. (2022). Reframing urban and regional 'development' for 'left behind' places. *Cambridge Journal of Regions, Economy, and Society, 15*, 39–56.
MacLeavy, J., & Jones, M. (2021). Brexit as Britain in decline and its crises (revisited). *The Political Quarterly, 92*(3), 444–452.
Marmot, M. (2022). The government's levelling up plan: A missed opportunity. *British Medical Journal*, 1–2. https://doi.org/10.1136/bmj.o356
Marmot, M., Allen, J., Boyce, T., Goldblatt, P., & Morrison, J. (2020). *Health equity in England: The Marmot review 10 years on*. London: Institute of Health Equity.
Martin, R., Gardiner, B., Pike, A., Sunley, P., & Tyler, P. (2021). *Levelling Up left behind places: The scale and nature of the economic and policy challenge*. Routledge.
Minton, A. (2019). Grenfell and the place of housing in modern life. In A. Whitworth (Ed.), *Towards a spatial social policy: Bridging the gap between geography and social policy*. Policy Press.
Mitchell, W., & Fazi, T. (2017). *Reclaiming the state: A progressive vision of sovereignty for a post-neoliberal world*. Pluto Press.
NAO. (2020). *The government's approach to test and trace in England—Interim report*. London: National Audit Office.
NAO. (2021). *The adult social care market in England*. London: National Audit Office.

Newman, I. (2014). *Reclaiming local democracy: A progressive future for local government*. Policy Press.
Newman, J., & Gilbert, N. (2022). The role of the private sector in subnational governance: Learning lessons from England's local enterprise partnerships. *Local Economy, 37*(1–2), 66–83.
Peck, J. (2003). Geography and public policy: Mapping the penal state. *Progress in Human Geography, 27*(2), 222–232.
Phillips, D., & Simpson, P. (2018). *Changes in councils' adult social care and overall service spending in England, 2009–10 to 2017–18*. London: IFS briefing note.
Pike, A., Coombes, M., O'Brien, P., & Tomaney, J. (2018). Austerity states, institutional dismantling and the governance of sub-national economic development: The demise of the regional development agencies in England. *Territory, Politics, Governance, 6*(1), 118–144.
Rhodes, R. (1997). *Understanding governance: Policy networks, governance, reflexivity and accountability*. Open University Press.
Rodriguez-Pose, A. (2018). The revenge of the places that don't matter. *Cambridge Journal of Regions, Economy and Society, 11*(1), 189–209.
Sandbu, M. (2020). *The economics of belonging*. Princeton University Press.
Skelcher, C. (2003). Beyond the sovereign council. Paper for the PSA Conference, Leicester.
Slater, T. (2018). The invention of the 'sink estate': Consequential categorization and the UK housing crisis. *The Sociological Review, 66*(4), 877–897.
Streeck, W. G. (2016). *How will capitalism end?* Verso.
Titmuss, R. (1968). *Commitment to welfare*. Allen & Unwin.
Tomaney, J. (2016). Limits of devolution: Localism, economics and post-democracy. *The Political Quarterly, 87*(4), 546–552.
Tomaney, J., Natarajan, L., & Sutcliffe-Braithwaite, F. (2021). *Sacriston: Towards a deeper understanding of place*. University College London.
Uberoi, E. (2021). *Turnout at elections*. London: House of Commons Library.
UK2070 Commission. (2020). *Go big. Go local: The UK2070 report on a new deal for levelling up the United Kingdom*. Nottingham: UK2070.
Warren, J., & Wistow, J. (2017). Policy, practice and difference within welfare regimes: Evidence from the UK. In B. Greve (Ed.), *Handbook of social policy evaluation*. Edward Elgar Publishing.
Wistow, J. (2022). *Social policy, political economy and the social contract*. Policy Press.

CHAPTER 5

Sentiments from a 'Left Behind' Place

Abstract Offering an empirical case study of 'left behind' Redcar & Cleveland in Teesside, this chapter explores the core problems in the area. Beginning with a discussion of deindustrialisation and an increase in insecure forms of work under neoliberalism, it then outlines how this economic restructuring contributed to the development of cultural problems. These include criminal activity like drug dealing and misuse, acquisitive crime including shoplifting as well as socially corrosive forms of anti-social behaviour. The chapter also explores the North South divide, as well as the residents' sentiments on the Levelling Up agenda and how they believe it can be a success.

Keywords 'Left behind' • Deindustrialisation • Localised problems • Crime

Introduction

So far, this book has provided a complex systems framing of the Levelling Up agenda through exploring and contextualising it from different entry points, including the trajectory of the political economy, implications for spatial inequalities, and changing dynamics in policy and governance systems, as well as analysis of capitalist trajectories and the implications for place. This chapter builds upon this, offering a case study of a 'left behind'

L. Telford, J. Wistow, *Levelling Up the UK Economy*,
https://doi.org/10.1007/978-3-031-17507-7_5

place and further exposing the challenges for the Levelling Up strategy. It thus explores the panoply of problems in the second most deprived 'left behind' local authority—R&C—as identified by Martin et al. (2021). It begins by documenting how these problems are longstanding, particularly regarding industrial work's previous centrality to the local economy. The deindustrialisation process meant stable and secure lifeworlds disintegrated, with myriad implications for many 'left behind' locales (Sandbu, 2020). This involved the intensification of poorly paid and insecure forms of work, coupled with the gradual decline of community spirit over the past forty years. As market values—competition, individualism, and primacy to the 'I' over the 'We' became hegemonic—the sense of localised camaraderie and relative collegiality that once characterised working class life faded from view (Lloyd, 2018; Winlow & Hall, 2013). Such socioeconomic decline has filtered into gradually degenerating local high streets. Whilst there was a degree of spatial disparity across the local authority, there was a perception that they were characterised by fast food outlets, charity and pound shops as well as boarded up stores.

In these zones of 'permanent recession' (Hall et al., 2008), various cultural problems are prevalent as criminal activity particularly volume crime like shoplifting, illicit drug dealing, and usage as well as violence are often normalized (Ellis, 2019; Kotze, 2019; Treadwell et al., 2020; Winlow & Hall, 2006, 2013). As we will see, corrosive forms of ASB are also prevalent in this 'left behind' place which often causes myriad distress to residents. Debates over a 'North South divide' have been prominent throughout neoliberalism (MacLeod & Jones, 2018; Martin, 1988), which was also reflected in our research as most respondents believe parts of London and the South are more prosperous particularly regarding the enhanced availability of better paid jobs and more state investment. However, there is a degree of spatial variation as no matter where you go in the UK—places of wealth often sit side by side pockets of deprivation including within this 'left behind' place (Boswell et al., 2022; Jones, 2019; MacKinnon et al., 2022; Wistow, 2022). As previously outlined, the Levelling Up strategy has proposed to remedy the UK's place-based inequalities (HM Government, 2022), but the chapter closes by documenting how there is ambiguity surrounding the agenda as it often represents different issues to residents.

Industrialised Security to Instability

As Teesside was central to the nation's economic growth and industrial expansion after the social and economic hardships of World War Two (Warren, 2018), a degree of stability and certainty began to characterise working class life in industrialised places in the sub-region during capitalism's post-war phase. One of the key findings from our research was a palpable sense that working class life seemed to work more easily decades ago. People spoke about how it was once relatively easy to find work, suggesting they could leave one employer and join another one within a week or two. This was particularly aided by the expansion of ICI in the sub-region, including the formation of another site in the 1950s. Innovating and leading the way globally in chemical production, ICI was recalled fondly by all participants. Charlotte, in her late 60s, suggested:

> Nearly everybody worked at ICI. Most people knew of somebody who worked there. My brother did, my dad did. On that estate up there houses were allocated to ICI people. A friend of ours moved from Scotland because he got a job at ICI. Houses were put aside for them. This town developed as almost a commuter town for ICI. My Dad used to go on a bike down Wilton to get there. It was on your doorstep. It was a good firm to work for and they looked after you. If you have an ICI pension; you are laughing. They looked after you so well. I can remember when I was in hospital as a kid—my Mam and Dad didn't drive so ICI paid their bus fares to come and see me. They used to say, you are made for life if you work for ICI.

The connection that many people across the subregion still feel towards ICI is palpable. ICI was stitched into the biographies of many of the area's families, since people would leave school safe in the knowledge that they could potentially acquire stable and secure employment with ICI. Equipping their workforces with comprehensive skills training and what was generally regarded as a 'job for life', some people moved from afar to take advantage of the company's remuneration and dependable working conditions. As the data outlines, housing estates and localities developed in light of ICI, illuminating how a place's *existence* was shaped by local industry. There was also a sense that ICI cared not only for its workers but for their families too. With remunerative jobs close by, people did not have to worry about lengthy commutes to work, often walking or in the case of Charlotte's father biking to work. This fomented an attachment to place, with *localised predictability* forming the basis of social life under capitalism's post-war phase (Hoggart, 1957). Such a dynamic was elucidated by Mark [retired]:

> I left school and got an apprenticeship as an electrician at Dorman Long steelworks and then I went into Smiths Dock shipyard. Dorman Long were offering me slightly more than Smiths Dock, so I ended up going back there. I worked there for over five years, and I went contracting. I never had any problems getting work. ICI was a brilliant employer for Teesside, you could get a job fairly easily at all levels. You could work your way up to the top. I know some lads who were not the brightest but worked their way up.

Regardless of one's educational achievements, working class people could leave school and obtain economically lucrative industrial work. Learning one's trade in an area of industry like electrical works also aided other employment opportunities, since workers possessed a skillset that was needed across the labour market. As Mark reveals, ICI existed alongside the steelworks and shipyards, with both renowned for their family-oriented working ethos and extensive training packages (Williamson, 2012). This developed people into dexterous industrial workers, involving opportunities for career progression and thus chances to scale up the institutional hierarchy. As mentioned in Chap. 2, in ICI's heyday in the 1970s it employed around 35,000 people, contributing to 43% of Teesside's residents employed in industrial jobs by the mid-1970s (Telford, 2022a). Full employment was achieved in parts of the area, with Anne [retired] claiming: "You could easily get a job in the 60s, 70s". Importantly, these were generally not degrading and poorly paid forms of work but jobs that equipped workers with a stake in both the present and the future. Employees could provide for their family, pay the bills, and have enough disposable income left for various leisure activities. Whilst this provided economic stability, the workplace cultures that often developed under capitalism's post-war era were generally characterised by reciprocal and friendly relations. Emma is retired and outlines how:

> It was hard work, but fabulous people who to this day we are friends. Lasting friendships and good times. It was like a community; everybody was there for one another. We had some fun times. It was very sad when it closed.

As Emma mentions, the heady days of industrialism in the area did not continue indefinitely. With post-war social democratic capitalism falling into a state of crisis in the mid-1970s, the trajectory of history changed and neoliberal ideals were enacted. As our book has outlined, this involved primacy to globalised market forces with little consideration of the myriad uneven spatial development it created in its wake (Harvey, 2005, 2007; Martin et al., 2021; Streeck, 2016; Wistow, 2022). Branch plants were continually shifted abroad throughout neoliberalism, meaning some places particularly across Northern England had their economic rug pulled from them rather quickly. Industrialised stability and security diminished from much of Teesside, unfolding in various waves including the industrial retrenchment across the 1980s under the Thatcher Governments; the gradual loss of ICI in the 1990s; and the closure of the steelworks in 2015. All respondents spoke about the detrimental impact of these changes upon the local economy, with the onset of a service economy not being an adequate replacement for the haemorrhaging of industrial work (Martin et al., 2021; Sandbu, 2020). Alan is retired and previously worked as a nurse. He said:

> Since they [industries] have closed, there aren't really many job opportunities around here now. There is Boulby not far from here and Skinningrove have a steelworks but it isn't very big. I feel sorry for younger people, as I do not know what they are going to do.

Respondents tended to view employment opportunities in both their town and across Teesside as inadequate, particularly in terms of well-paid work. As Alan mentions, there is still some industry in the region although it pales in comparison to the area's former industrial prowess. Boulby mine opened in Teesside in 1973 and was the first potash mine in the country (Telford, 2022a), though it now produces polyhalite. Whilst the mine previously employed over 1000 workers, it has gradually downsized its workforce and now employs around 450 people. Moreover, Skinningrove steelworks dates as far back as 1880, though in more recent decades it has encountered various rounds of privatisation and redundancies. As of 2021, the plant employed 327 people and produces steel for companies around the world like JCB and Toyota (Hughes, 2021). However, as Stuart in his late 40s highlights: "The area has lost a lot of better paid and skilled jobs", meaning what remains industrially is not sufficient to meet the demand for remunerative work. As Alan notes above, this is particularly damaging for younger people who have grown up in the deindustrialised era. This includes Danny, 21. He previously left college and applied for various apprenticeships, eventually being offered one in engineering. He had this to say:

> We have the Wilton site where there is a bit of industry and good companies. They are good jobs. Apart from that, it isn't great for opportunities. We lost the steelworks, which was bad. It is mostly retail and shit like that.

Although ICI moved parts of its plants offshore and privatised what remained, Teesside still possesses the second largest chemical complex in Europe in terms of manufacturing capacity. The region's chemical industry has three core sites across Teesside at Wilton, Billingham and Seal Sands which contains international businesses like Sabic and Huntsman (Tees Valley Mayor, n.d.). These jobs are highly skilled and well-paid, though there was a perception in our sample that it is very competitive and thus difficult to 'get into' the industry. Whilst these employment opportunities are at a premium, other segments of the local labour market are characterised by poorly paid and non-unionised employment. The loss of the steelworks in 2015, which resulted in over 2000 redundancies as well as many more in the local supply chain, accelerated this trend. Whilst most of the steelworkers acquired work in the subsequent years, it tended to be in lesser paid jobs (Telford, 2022a). As Danny mentions, there are plenty of jobs available in sectors like retail, though these are generally minimum wage, degrading and often defined by a high turnover of staff (Lloyd, 2013, 2018). These forms of work generally fail to equip workers with a stake in the present and the future and offer little sense of pride and social purpose. Such degradation of employment conditions was illuminated by David, retired:

> A lot of the happiness has gone out of jobs—too much focus on managers, budgets, shareholders; money for the boys.

The debasement of work particularly in 'left behind' places throughout neoliberalism means low-paid work is often degrading and fails to provide job satisfaction. There was often a sense that there was too much emphasis upon profitability and efficiency savings at the expense of the workforces' working conditions and wellbeing. Many spoke about how pay has not kept up with the cost of living particularly in recent years, with inflation reaching 9% in the summer of 2022 and thus its highest level in several decades (Etherington et al., 2022). Whilst the UK Government have emphasised the buoyancy of the labour market and how labour shortages in certain sectors are contributing to rising inflation, for the respondents these notions were rather simplistic. This includes Gareth, in his mid-50s:

> I am not sure if there are a lack of jobs, but a lack of good jobs. I think people can find work, but the question is can they get paid well? The principle of going to work is to better yourself; get a salary where you go do well and provide for your family. That principle has been *left behind* somewhat—look at all the people in work on universal credit.

Other commentators have also highlighted how in many 'left behind' places low-paid work means people often must rely upon the crumbling welfare state to try and make ends meet (Beatty et al., 2022; Etherington et al., 2022). Statistics, for instance, indicate that the average full-time workers' gross weekly pay in R&C in 2021 stood at £511 in comparison to both the Northeast and UK average of £547 and £613 respectively (ONS, 2021a). Indeed, the structural combination of deindustrialisation, a dearth of remunerative work and the advancement of low-paid employment under neoliberalism has impacted detrimentally upon the broader locality, particularly via the decline of a sense of community and the high-streets.

"Everything is Run Down"

Whilst the industrialised era fostered a sense of community and a degree of reciprocal cultural relations (Hoggart, 1957), most of the respondents claimed these characteristics were absent from their localities today. Most participants emphasised how socio-cultural life had been declining for many decades, with a sense of civility and neighbourliness fraying at the seams. Emma said:

> I grew up in an area where everybody knew each other, which is great as you never know when you need people to help you. When I was growing up, that's what it was all about. Very close-knit, totally different to now. You used to feel safe; we never locked our door. Even my dad up until he died 20 years ago, he never locked his door. You could never do that now. I am thankful that I grew up then, rather than growing up in today's world. It was a much simpler and happier life. We were always involved in local activities. I wouldn't feel safe going out around here now on a night. There isn't

> a lot of good these days; there is a lot of sadness. The person who used to live next door to me, I got them a welcome card, said hello. But they would only answer me, they wouldn't make conversation. People used to be more caring, more interested in other people. People used to take more pride. It might sound idyllic, but to me it was.

Emma describes a socio-cultural landscape that has changed enormously since the neoliberal 'phase shift' in the political economy, fostering localised feelings of insecurity and uncertainty. As we will see, this is aggravated by the prevalence of crime and ASB in the local authority. Particularly the older residents emphasised how social values like reciprocity, obligation, and relative commitment to one another had faded away, replaced by social atomisation and individualism. This further illuminates how debates on 'left behind' locales need to be about more than their economic decline, and to take into account issues such as the importance of more traditional social values (Boswell et al., 2022; Goodwin & Heath, 2016). Nonetheless, the degradation of community life in the wake of deindustrialisation under neoliberalism has been noted by other scholars (Lloyd et al., 2021; Telford et al., 2022; Winlow & Hall, 2013). Anne, retired, previously worked in Teesside's textiles industry. She also expressed how community life had been diminishing:

> *We have no community*. I was going to do a jubilee party in a couple of weeks, and nobody wanted to go. I was going to do it in the square, but nobody was interested. There was a better community years ago. We had lovely neighbours; my parents used to socialise with them all the time.

A sense of commonality and community that once characterised many formerly industrialised but now 'left behind' places have dwindled. This is particularly the case in areas like R&C, since industrialised places tended to be spatially organised around housing that was 'mostly terraced with low garden fences', 'the bells and whistles of work time, the joint movement of bodies through the streets at this time, the sounds of the works themselves' (Walkerdine, 2020, p. 152). The absence of this mode of work and life has had feedback into the spatial dynamics of place as a form of sociality in localities like this has diminished. In fact, post-social arrangements are often present (Winlow & Hall, 2013), that is, a sense of community and looking out for one another has declined while individualism and communal fragmentation are predominant. As Alan stated: "When I was younger, you wouldn't lock your door. Now, I lock the door even when I am in the house." The perceptions above regarding a *lost community* were also related to the decline of the high streets. Most of the sample suggested a once vibrant social space had been deteriorating for several decades, particularly for those who lived in the local authority's market towns. Laura, in her 60s, has worked in minimum wage jobs throughout her working life including as a dinner lady and cleaner. Of the sample of interviewees, she lives in one of the most deprived neighbourhoods and said:

> All the shops have closed and are boarded up, so you can't just pop to the shop. If you haven't got a car, it is hard for them [pensioners] to get anywhere... The high street is just all charity shops. Years ago, the market was absolutely brilliant, you could spend all afternoon there. It is nothing like it was. A lot of these market towns aren't like they were.

The high street's decline was spoken about with sadness and resignation, shaping the localised 'feeling of left behindness' (MacKinnon et al., 2022, p. 44). As our data indicates, some of the local authority's high streets are populated with closed and empty stores, as well as charity shops. While some respondents spoke about often acquiring some 'bargain' items in these shops, Margaret [retired, previously worked as a midwife] suggested: "The amount of charity shops gives you an idea about the problems around here". Indeed, more deprived regions often possess more charity shops in their high streets. Statista (2019), for instance, highlighted how the lowest proportion of charity shops were in Yorkshire and London at 4.4% and 5.8% respectively, while the North East's national share stood at 8.9%. This filters into a general sense that the high-street offers little social substance, as Danny explains:

> There is not a great deal there [high street]. Lots of pizza shops, takeaways. Wetherspoons opening has put a lot of pubs in difficulty, especially with COVID, but 'spoons' is cheaper. We have had quite a lot of shops closed that have then changed into takeaways or phone shops.

The local authority area has two Wetherspoons pubs, with one opening in 2017. As Danny indicates, the cheapness of 'spoons' tends to impact detrimentally upon other pubs in the area since they cannot financially compete with their prices. Danny also alludes to how the economic difficulties of the high street was amplified through the COVID-19 pandemic, particularly as trade was halted for prolonged periods during the lockdowns. Employment opportunities at Wetherspoons also filters into the local authority's lack of remunerative jobs, with research indicating that Wetherspoons employment tends to be poorly paid, non-unionised and defined by a high turnover of staff (Cant & Woodcock, 2020). As mentioned in the previous chapters, the HM Government (2022) highlights the importance of 'restoring a sense of community, local pride and belonging' particularly by rejuvenating 'left behind' place's town centres by 2030 as part of the agenda's missions and objectives. But at the micro level of interaction the breadth and depth of the problem in places like this is such that sustained investment and support is required to tackle both local authority area *and* within local authority spatial variations that contribute to constituting the 'left behind' problem. Therefore, it is important that some policy measures are tailored to the specifics of place, which was illuminated by some respondents who often highlighted how some high streets in the area are more vibrant than others. This includes Stuart:

> This high street here is doing well in comparison to some high streets in the area. Some of the places not far from here; it is like the third world. Total lack of shops, *everything is run down.*

Such spatial variation is important, particularly as spatial inequalities are prevalent within regions whereby 'prosperous communities are found cheek by jowl with areas of social and economic deprivation' (Chaytor et al., 2020, p. 11). Stuart later indicated that he was referring to towns and villages in parts of rural East Cleveland; many of whom historically depended upon the extraction of iron-ore and potash mining (Kalantaridis, 2010) but have been economically abandoned in the neoliberal era. East Cleveland has generally contained higher than national average levels of unemployment for several decades, involving the outward migration of some of its younger residents who leave to search for a better life elsewhere (Kalantaridis, 2010). This spatial variation regarding the local authority's decline was illuminated by Gareth:

> A lot of places are out of sight and out of mind. Drive through Ellers, Campion, Bartley, Lakeside and Bosworth and you think what on earth have these people got? What is there for them? These places absolutely need some investment—desperately so. A lot of their high streets are basically derelict,

eyesores. They don't get any money spent on them. Most of the money gets spent on HillBrook, and the rest are just left to pick up what little money is left. Eventually people just want to move away from these places, just like the seasonal coastal towns. A lot of them have just become *ghost towns*. Some of their high streets, there isn't much going on at all—charity shops, take-aways, more pizzas than you can eat.

As the data illustrates, many 'left behind' localities have been economically abandoned throughout neoliberalism, suffering from an inability 'to adapt to the growth of the post-industrial service and knowledge-based economy whose locational requirements are very different from those of past heavy industries' (Martin et al., 2021, p. 108). Localised characteristics including their high streets thus decline, partially since many residents do not have the disposable income to spend in their community. Gareth's perception that most of the local council's money is also spent on neighbouring HillBrook, which has particularly suffered from the loss of ICI and the steelworks over the past several decades, was also reiterated by most respondents. Such spatial divergence means some people in 'left behind' localities desire to move away in search of a better life, with Danny stating: "It [the area] feels like home. I don't really want to move away. But if I have to—I will." This sentiment is brought by a dearth of lucrative jobs, declining localised conditions like a sense of community, degeneration of the high-street and the culmination of some nearby locales into *ghost towns*. The prevalence of fast-food stores was also noted across the sample, with research indicating that there is 'a clustering of these outlets in more deprived areas' (Lake, 2018, p. 244), often impacting negatively on the local population's health. This is further impacted by what Anne claims is a:

Lack of transport. They took the bus away that used to come down here to take us up street. I used to get on it a lot to socialise, but I can't now. They stopped it at the beginning of lockdown, and the bus hasn't been put back on. It is a Tees flexi bus, but it doesn't run as much now. The shops on the high-street is full of charity shops and pubs. It has changed a lot. Where are the decent shops? There's also a lack of facilities for the kids, there's nowhere for them to go so they cause havoc.

As we previously discussed, austerity measures weakened social infrastructure particularly in 'left behind' areas across the North (Gray & Barford, 2018) and Anne alludes to how this was aggravated by the

COVID-19 lockdowns. This socio-cultural decline also effects younger people, since they often possess little sense of both social purpose and a future in 'left behind' places. All respondents claimed the above issues were contributing to crime and ASB in the area. These were core local concerns and are presented as the central topics of discussion in the next section.

Crime and Anti-social Behaviour

When the participants were asked about the key problems in the local authority, crime and ASB were often their first responses. Particular forms of crime included vandalism and the prevalence of illegal drug dealing and consumption, as well as acquisitive forms of crime like shop lifting. Others spoke about how ASB involved verbal abuse, fly-tipping, graffitiing often offensive terms onto public buildings and pavements, littering, coupled with a general decline in civil manners in recent years. All respondents spoke about how these issues were serious problems and socially corrosive, often generating psychological distress. Whilst these issues have not featured much in the 'left behind' and Levelling Up literature, as mentioned reducing neighbourhood crime forms one of the HM Government's (2022) key missions and objectives. Stuart claimed:

> More people think drug use is acceptable these days, go back a few years ago and it wasn't. The lad in the upstairs flat is addicted to crack cocaine. He worked with me and lost his job because of it and has lost his family…. It worries me that a lot of people get into drugs because there is *nothing for them around here*. When we were younger, we had things to do and look forward to, we had something to do. We had a focus; it is a big issue.

The prevalence of illicit drug use particularly in the more deprived spaces formed a severe blight upon their neighbourhoods, especially the consumption of heroin and crack cocaine. Such drugs generally amplify peoples' personal problems and distress in the long-term, ensnaring them in a cycle of addiction and acquisitive crime particularly in the context of neoliberalism and drug prohibition (Wakeman, 2021). The links between socio-economic deprivation, unemployment, impoverishment, and higher rates of drug addiction are well established (Ayres, 2021; Coomber & Moyle, 2018; MacDonald et al., 2020). For instance, Middlesbrough, which borders parts of R&C, possesses some of the highest rates of heroin use in Britain and has been branded as the 'drugs capital' (BBC, 2021). Contextual conditions in the area means some people in 'left behind' places turn to drugs to temporarily escape a difficult reality, including some young people who use them recreationally:

> There's a lot of drugs these days. You see them at the park smoking them long fags; sit in the shelter doing them. You can smell it when you go by, they are only about 11. Where do they get it from? A lot of kids put windows through. They carry on in shops like the supermarkets. The other day I saw a bloke with a rucksack coming out and the security man was chasing him down the high street. [Laura]

Similarly, Jodi is in her 40s and works as a hairdresser earning the minimum wage. When asked about the area's problems, she said:

> Lots of drugs. Police were driving downtown when I was walking home, the druggies had been shoplifting in town. I know people who work in supermarkets around here; they said it happens all the time. But their policy is not to approach them… There was a stabbing the other week in broad daylight, all sorts of weird things. When I walk to work, there is a flat I go past, and the smell of weed is so strong. They are growing it and selling it there; it absolutely stinks. Really strong smell.

Crime is something that these localities endure as a frequent and socially harmful force, whereas crime for those living in the more affluent neighbourhoods of the UK is something distant; only encountered through the news or shows on television. As the data indicates, drug dealing and misuse has recently been accompanied by a stabbing in the area, which most respondents highlighted. Although many were taken aback when it initially happened, they suggested it is not a surprise as crime had been getting worse in the area for some time. Laura noted how:

> There was the stabbing the other day. Somebody was getting their money out of the cash point, and someone stabbed them in the leg with a vegetable knife. It was to do with drugs; the lad owed him money. It was in the middle of the day. Old people were doing their shopping. Our bairn saw it as she was walking home from school. Around the back of ours the other day, one house got raided. There was five police vans and two police cars. It was about drugs, I think. People have been setting their bushes on fire, smashing their windows.

The prevalence of illicit drugs acts as a corrosive cultural undercurrent in such areas of 'permanent recession' (Ellis, 2019; Hall et al., 2008), whereby 'criminal enterprise has become an everyday fact of life for many' (Treadwell et al., 2020, p. 59). Indeed, ONS (2021b) data indicates that Cleveland police force—which also covers other deprived areas like Hartlepool across Teesside—has the worst recorded levels of crime in the country. It is worth mentioning that recorded crime only indicates the tip of the iceberg as the 'dark figure' of criminality is unknown (see Kotze, 2019). This was illuminated in our study as despite some of the respondents regularly witnessing illicit drug dealing and usage, they never reported it to the police. Bryony is retired and spent her working life as a police officer. She said:

> I know of a few break ins that have happened at peoples' garages; it happens a lot of places now. They bring in people to these estates from very rough areas. When you bring them in, it is game over. Around the park, there are kids hanging around and I just don't go in as they are intimidating. They damaged allotments not long ago; they killed the chickens and pigeons. The mentality of these people. They don't care. They couldn't care about the police or other people.... It is like the other day when that shop window was put through at 1am in morning. I didn't hear it but the sirens flashing woke me up. They were effing and blinding at the police, kicking off. I was thinking, for God's sake just get rid of her and put her in the van.

It is difficult to understate the blight that crime—or what Hall and Winlow (2018, p. 113) refer to as 'Little Evils'—have upon 'left behind' communities. Exacerbating post-industrial locale's atmosphere of uncertainty and instability, they corrode socio-cultural relations not least as crime tends to be intraclass (Ellis, 2019). They also amplify the localised sense of fear and uneasiness, with crime both normalised and deeply embedded in the locality. Bryony also highlights a sense of corrosive individualism and social atomisation that has replaced the once brittle but relatively communal bonds in these working-class places. Such sentiments were illuminated by all respondents including David: "Manners and anti-social behaviour have deteriorated all over". As the data indicates, high-crime locales like this contain different forms of criminality that often cause myriad harm to the local population, which tends to be most acute in deprived neighbourhoods. However, it is important to point out that criminality is a complex symptom of structural causes, with deindustrialisation, poorly paid work and the degeneration of the community and high streets forming the contextual background in this study. Nonetheless, Charlotte claimed that the pervasiveness of crime means that:

> Nearly everybody has a camera on their house now. You never used to hear of that. I am in between two people who have cameras; I want one fixing on mine. A person on their own—I need one and would feel better. Before I go to bed, I always check my car. If I wake up in the middle of the night, I check my car. I don't want it damaging by people; terrible anti-social behaviour.

Evidently, crime is a socially menacing force in 'left behind' places like this which impacts negatively upon many aspects of peoples' lives. For some residents, this even includes encroaching upon their sleep since they are bedevilled by a sense of fear about being a victim of criminality and ASB. Such blight upon certain neighbourhoods perhaps explains their support for tough on crime policies, with the HM Government (2022) promising to cut crime and ASB in high-crime places by 2030. Indeed, the growth of situational crime prevention measures like private security cameras also forms part of the broader securitisation of society under post-social neoliberalism, amplifying a sense of fear and suspicion of other people (Raymen, 2016). People are viewed as a potential social threat and risk rather than a source of sociability and cultural enrichment. Private security cameras also transfer the responsibility for rooting out crime and ASB away from the state and criminal justice agencies like the police onto communities, reflecting the broader shift towards individual responsibility under neoliberalism (Kotze, 2019; Raymen, 2016). Such security measures were also enacted by many participants partially because of a lack of faith in the police to do their job. While some respondents situated their discontent with the police in light of austerity and the loss of police officers and resources, it did not diminish their frustration:

> The police just drive up and down in vans. They used to walk around more. They used to be able to get to know people; build relationships with the younger ones. But they don't do anything anymore. Even them street wardens; you don't see them. A lot of the younger ones drive around on motorbikes—I have seen them fly around here with no helmets on. The police don't do anything. [Alan]

Evidently, cuts to the police have had a detrimental impact on the ground, involving a lack of resources to respond to crime and ASB. However, Cleveland Police was put in special measures in 2019 after being rated inadequate on several levels, following an inspection by Her Majesty's Inspectorate of Constabulary and Fire and Rescue Services (HMICFRS). The inadequate rating covered issues including: preventing crime and tackling ASB; protecting vulnerable people; fair treatment of the public; and ethical and lawful workforce behaviour (HMICFRS, 2019). The force was deemed particularly inadequate in responding to less serious forms of crime, which were not always allocated to adequately trained and experienced staff and not investigated thoroughly enough (HMICFRS, 2019). While improvements have been made, the force is still in special measures over three years later. Such problems meant that people were often surprised when the police responded to crime in the area. Jo, in her 50s and a dinner lady of over 20 years, highlighted how:

> The house around the corner got their windows put through not long ago, as well as another shop window over there. It was good to see the police actually come out, as they usually don't.

The general lack of response from the local police force amplified peoples' sense of feeling 'left behind'. This was reflected in Stuart's statement regarding the police's lack of investigation into a crime in a different part of R&C where he lived two years ago: "The police were useless; they had almost written the ward off". Deindustrialisation, the advance of insecure and poorly paid forms of work, the decline of the local high streets, a loss of a sense of community, as well as the prevalence of crime and ASB filtered into a general perception of a 'North South divide'. This involved how the North had been declining for several decades, while London in particular had pulled ahead of the rest of the nation. This issue is where our book now turns.

"We are a Forgotten People"

As mentioned in Chap. 2, in light of deindustrialisation's advance and the Big Bang deregulation of the British economy in the 1980s a debate emerged about a 'North South divide' (Jones, 2019; MacLeod & Jones, 2018; Martin, 1988; Wistow, 2022). This idea has been a stable feature of neoliberalism, central to both the UK's position as one of the most regionally unequal nations in the developed world and thus the Levelling Up strategy (Martin et al., 2021). The HM Government (2022, p. 5) highlighted how London now forms one of many of the world's 'knowledge-intensive super cities' alongside San Francisco, New York, and Tokyo, resulting in other places falling behind. Whilst most respondents were keen to emphasise R&C's natural beauty due to its proximity to the countryside, existing on the cusp of the North Yorkshire moors, as well as its closeness to the coastline, there was a perception that the North South divide had been congealing for many decades and was so all-encompassing that it could not be adequately bridged. When asked about this divide, Anne said:

> Money wise, everything, facilities. Go down South and it is a different world. It gets way more money down there; more spent on it. We pay all that council tax and get the least spent on us. I think it goes further back than the cuts that have happened—it is a big North South divide. They all think we are flat caps and whippets; they look down on us. Put it this way, it goes as far up as Manchester, then the rest of the North is left. It is as far North as they go. Even the weather, when they show you the weather they show you Manchester, what about us up here? We are a *forgotten people* up here.

The subjective sense of a North South divide 'left-behindness' is largely corroborated in Martin et al.'s (2021) study of 'left-behind' places (as measured through GVA and employment over an approximately 40-year period), which also covers important caveats about spatial variation within regions and localities. Higher levels of public investment in London have been a longstanding feature and were intensified in the austerity age. As the Institute for Public Policy Research (IPPR) (2022) indicated, in the five years leading up to 2019/2020 London received £12,148 per person while the North received £8125 respectively, meaning if the North received London-level investment it would have obtained £61.6 billion more across the five years. As mentioned, austerity was also not felt evenly, serving to compound this regional spatial imbalance. Areas across the

Northeast were hit the hardest through regressive cuts to the public sector, including to local authority budgets and the closure of many public services (Bambra & Garthwaite, 2015; Beatty & Fothergill, 2018; Gray & Barford, 2018). As the data highlights though, the North South divide predates austerity with locales like R&C scarred by a legacy of industrial retrenchment, meaning deindustrialisation and austerity formed a toxic symbiosis in the area. This combined to foster a localised feeling that places like R&C were discarded in the neoliberal era. Indeed, Bryony purported that:

> I have friends in the South, and they haven't got a clue about life up here. You go down to London, transport is brilliant—buses, tubes, trains. The transport up here is terrible. There is a big divide. Jobs wise it is a different world down there. I know people who work in the city—very good jobs. The capital has pulled away of everyone else—but I suppose some cities are different like Manchester and Leeds. We lost a lot of industry up here. I was brought up with cotton mills, mining, but it all went abroad as it was cheaper. The area needs its socks pulling up. I hate some of the areas around here—River Way, HighBrook.

For some respondents, there was a public disconnect regarding the different material conditions in areas like 'left behind' R&C compared to parts of the South. This divide cuts across numerous axes including investment, public infrastructure, and the availability of remunerative jobs. As Martin et al. (2021) highlighted, London has seen an explosion in high-paid wealth, finance, and professional business-based employment. For instance, it has around 700,000 people who are employed in banking and finance (Atkinson, 2020); industries which serve to extract wealth from the British economy and intensify inequalities (Wistow, 2022). It is no surprise that the North received £515 less per person than the nation's capital in transport spending over the last decade (IPPR, 2022), amplifying unequal spatial disparities. Such divides were also somewhat reflected by Danny: "The South is closer to the capital; it gets a lot of tourism and gets more heavy investment."

It is worth reiterating that the idea of a North South divide fails to capture the complexity of the UK's place-based inequalities. No matter where you are in the UK, social and economic deprivation often sits side by side with pockets of affluence (Boswell et al., 2022; Chaytor et al., 2020; IPPR, 2022; Jones, 2019; Martin, 2021; Wistow, 2022). Whilst

London is generally thought of as a place of affluence, the capital possesses the highest concentrated poverty rate in Britain (IPPR, 2022), involving around a quarter of residents living in poverty (Atkinson, 2020). The city is partially characterised by spiralling inequalities, gentrified working-class neighbourhoods, homelessness, evictions and rising private rents meaning 'the London of today takes on the excesses and poverty of its Edwardian and Victorian counterpart but without the sense of social mission of those times (Atkinson, 2020, p. 167). Jodi continued the story:

> In central London, there is a lot of wealth but there is also a lot of poverty. My daughter works down there and gets more wages than up here, but what they have to pay for rent and the cost of living is far more than up here. Wherever you go, there is lower class, working class, middle class, and upper class. You get good and bad places all over the world.

Whilst the average workplace weekly earnings in London is the highest in Britain at £751 (Etherington et al., 2022), for many working-class people their wages are offset by the higher than national average cost of living in the nation's capital. For example, the average cost of a house in London reached over £500,000 for the first time in 2021 (Osborne, 2021). Therefore, as Jodi highlights, arguably the key divide across Britain's spatially imbalanced nation is social class. This was illuminated by many respondents like Margaret: "I can remember going on a trip down South and the area was really bad. We were told at the hotel to not go out on our own. You still get areas of poverty down South, and you still get areas of affluence around here." Therefore, the notion of a North-South divide is based upon the premise that the North is homogenous (IPPR, 2022), which glosses over how there are spaces of prosperity in R&C and across Teesside. David outlined how:

> As we are in TS14, we come up with bad results on things as we got bunched with all that lot in River Way—full of criminals. How can you compare Driers Wood to Canon Way? They are worlds apart, but in the same postcode.

It is important to not mask the nuance of the UK's geographical inequalities especially within regions and communities (Boswell et al., 2022; Chaytor et al., 2020). As our data refers to, there are some neighbourhoods in R&C where the houses are worth over £1 million while others are characterised by higher than national average levels of crime, drug addiction, poverty, poor housing stock and unemployment. This was also reiterated by Greg, in his 60s: "You get your bad areas and good areas wherever you go". Similarly, Stuart said:

> You can divide things up how you like. You can go ten miles from here and some of the houses are nearly a million pound. Look East to West—look at Blackpool and parts of Cumbria. Some of the towns there it is like *stepping back in time*. They look run down and dirty, especially if you are out of the tourist places. I avoid London like the plague—it doesn't interest me at all. The big divide for me is between those that have money and those that don't.

The UK's geographical inequalities are deep-rooted and pervasive, spanning North, East, South and West. For instance, Blackpool in the Northwest was a relatively prosperous coastal and tourist locality including under capitalism's post-war phase. But like many coastal places has been 'left behind' under neoliberalism and witnessed the emergence of higher than national average levels of joblessness, crime, poor housing stock and precarious work (Telford, 2022b). As many participants noted, the key antagonism is social class and economic inequality. Whilst more unequal societies such as Britain produce poorer health and social outcomes across mental health and wellbeing, education, trust, violence, crime, drug misuse and work (Pickett & Wilkinson, 2010), tackling economic inequality has featured little in the Levelling Up debate. This is particularly important as: "some of our historic local problems that we have seen for years are becoming more national problems. Just look at the *cost-of-living crisis*, it isn't confined to Teesside but is up and down the country [Gareth]." This ongoing crisis combined with all the issues above makes Levelling Up both urgent and essential to the redevelopment of places like R&C, and the respondents' sentiments about it are explored in the next section.

Levelling Up?

As Chap. 3 outlined, the Levelling Up agenda has been a staple feature of British politics since the 2019 general election. Despite its political prevalence, most respondents possessed different interpretations over what it means in practice. However, a key finding was their belief in the importance of well-paid work to the strategy; work that would give residents an economic platform to harness a stable and secure livelihood. Mark suggested:

> I have heard about it [Levelling Up]; we will have to see if it works. I am a Northerner, and we need better transport, more jobs. If people can get good jobs, that makes a hell of a lot of difference. We need to make sure that people are always better off if they work.

The importance of employment to working class life cannot be overstated, with work forming the economic backbone of individuals' lives, families, and communities. As we have seen, once remunerative work diminished under neoliberalism and was not adequately replaced various socio-cultural problems emerged as places became 'left behind'. The renaissance of well-paid work is a key policy challenge for the Levelling Up agenda and social policy more broadly, with deindustrialised R&C arguably requiring a new raison d'être to flourish again. As the Conclusion outlines, this could involve being a key part of a Green Industrial Revolution. Nonetheless, as mentioned there is a sizable discrepancy in funding for transport between the North and London, involving particularly poor transport links in rural communities across R&C especially to the East which the Levelling Up agenda needs to address. While some respondents like Jodi claimed: "I've heard the politicians say that [Levelling Up] a lot, but I'm not sure what it means", others like David were aware of some of the Levelling Up initiatives going on in the local authority and were cautiously optimistic:

> I think at long last after all these years they [politicians] have realised we [R&C/Teesside] are good at doing trades. There are more people up here who are good at doing hard work. If it wasn't for us, we wouldn't have had the industry. They have moved some government bases here. There is the freeport, and we are on about manufacturing batteries for these new electric vehicles on Teesside. They are on about bringing all sorts in; there is talk of thousands of jobs. People will vote Tory if they can bring jobs.

Of course, whether the Levelling Up strategy was developed by the Conservative Government because they genuinely care for the people who inhabit 'left behind' places, or whether they are primarily motivated by electoral calculation and retaining seats across the fallen 'Red Wall', is up for debate (Hudson, 2022; Tomaney & Pike, 2020). Nonetheless, the agenda is perceived by some residents as a welcome boost to the area, particularly in light of long-term economic abandonment. As displayed though, the locale's former industrial prowess and how it led the way in manufacturing particularly through its chemical industries still lives on in the collective psyche particularly of the older residents. However, as Telford (2022a) outlined it seems fitting for the area to be at the core of capitalism's next phase—potentially, a Green Industrial Revolution and thus investment in renewable industries including manufacturing batteries for electric vehicles. As we will see, this potentially forms part of deglobalisation, the shortening of supply chains and investing in the UK's manufacturing base around green industries in a post-neoliberal world. As David and others mentioned, if the Conservative Government can deliver on jobs, then they are far more likely to retain support in these places at the next general election. Similarly, Greg suggested:

> They [Conservative Government] are trying, aren't they? Whether you like Boris Johnson or not, he has had a lot to deal with. Levelling Up isn't going to happen overnight. It will take at least four or so years. Government buildings have moved to Darlington, but I don't think that is going to make a lot of difference. It remains to be seen whether the freeport will amount to much.

Given the scale and depth of the structural problems in 'left behind' locales and how they have been congealing since the 1970s they will take sustained economic investment over a lengthy period to ameliorate (Hudson, 2022; IPPR, 2022; Martin, 2021; Martin et al., 2021; MacKinnon et al., 2022; Tomaney & Pike, 2020). As mentioned, this is somewhat reflected in the current Government's declarations about Levelling Up, with the HM Government (2022) pointing to 2030 as the year it hopes to achieve its goals. However, currently absent is the tectonic economic resources required to achieve the state's Levelling Up ambitions. Martin et al. (2021) noted how it possesses some parallels to the profoundly spatially imbalanced German economy in light of German reunification. Since the collapse of the Berlin Wall in 1989, Germany has spent around €2 trillion on its Aufbau Ost scheme which aims to Level Up parts of what was formerly 'East Germany' to 'West Germany's' living standards. Therefore, the scheme has been retained by successive Governments and equates to around £55 billion per annum (Martin et al., 2021), which is far more than has been allocated so far through the Levelling Up fund.

Interestingly, most respondents associated Levelling Up with Teesside's Freeport but were unsure about how it would specifically benefit the area. Opening in late 2021 as one of eight freeports across the UK, Teesside Freeport consists of 4500 acres involving old industrial sites like the demolished steelworks, seaport, and an airport (Hall et al., 2022). As Tim said: "I have seen them [workers] clear all the old steelworks site; they are building roads and the infrastructure for a freeport so we will have to see what happens there". Cast by Hall et al. (2022, p. 3) as 'neoliberal policy laboratories', these special economic zones enable wealthy people to store their valuable commodities and not adhere to the nation's regulations like tax arrangements. As Tomaney and Pike (2020) outlined, freeports are unlikely to have any significant impact upon addressing spatial imbalances in 'left behind' places. Furthermore, freeports including in Dubai and Geneva have been known to facilitate illicit trade and generate heavy pollution which contaminates water and air quality, while their deregulatory environment means that many of the jobs created are low-pay and non-unionized (Hall et al., 2022).

This policy has been accompanied by shifting some Government Department's offices to nearby Darlington in Teesside, including 200 staff from the Department for Culture, Media, and Sport with plans for many more by 2025. Alan offered his views on this:

> They have moved the Treasury thing to Darlington; at least that is something. Well, they have probably moved it because the rates to run it are far cheaper up here than London, so it is to get more money in their pockets. They [politicians] don't seem to be bothered about this area.

Hudson (2022) outlined how relocating Government offices to the North is rather tokenistic and will not do anything significant to solve place-based problems. These types of developments are also often met locally with cynicism, scepticism, and fatalism amongst some residents (Telford & Wistow, 2020; Warren, 2018). Such feelings are elucidated by Alan, suggesting the relocation of Governmental Departments to Teesside could be more about fiscal sensibility rather than a desire to improve peoples' livelihoods in 'left behind' locales. This distrust of politicians is deep-rooted (see Telford & Wistow, 2020; Telford, 2022a) and whilst the HM Government (2022) often emphasises the importance of strengthening trust to the Levelling Up strategy, it will take concrete improvements in peoples' lives for this to occur. The cynicism and mistrust in our sample often meant that a handful of respondents were profoundly disengaged from politics and did not know what Levelling Up was: "I haven't heard much about it… But things need to improve. Get the shops back open. Get the bus back on" [Laura].

Tim identified more historic place-based challenges to Levelling Up:

> How are we going to Level Up and be like London? Are we going to get the same wages and jobs like they do down there? I don't think the North will ever be Levelled Up to the South. There's always been a lot more people unemployed up here; companies have more choice of staff to take on. They aren't going to raise wages when they can just take other people on.

Whilst the participants' views on Levelling Up were multifaceted, all involved the importance of well-paid and stable work. The dearth of this in 'left behind' R&C and prevalence of it in more prosperous places was cast by many participants as the key driving force behind the nation's spatial unevenness. For Steve, this problem is coupled with the historically higher level of joblessness across the North/Northeast in comparison to other parts of the UK, with what is deemed a natural level of unemployment under neoliberal capitalism awarding employers the ability to pick and choose workforces more easily (Lloyd, 2018; Winlow & Hall, 2013; Wistow, 2022). As Chap. 2 outlined, policies were enacted in the 1980s that purposefully created a reserve army of labour to undermine trade unionism, with post-industrial and 'left behind' communities living with the consequences over forty years later. Whilst the UK Government have emphasised labour shortages as we emerge out of the COVID-19 pandemic and into a cost-of-living crisis, the key sentiments in our research were concerns around inadequately paid work rather than a sense that there were insufficient employment opportunities.

Many respondents argued that Levelling Up was about more than just economics with, for example, Emma, stating that:

> I also think we need a big cultural change. There has been a decline in caring for one another. Simply holding doors open for people, basic manners, please and thank you. I place great emphasis on the family, and family life has suffered with both parents needing to work. Materialistically we might be better off, but families just don't have the time anymore.

The advance of market values into every aspect of society and social relations under neoliberalism has meant competitive individualism has frayed civility and sociality (Telford et al., 2022; Winlow & Hall, 2013). Such liberalisation of social relations alongside changing labour market dynamics under neoliberalism have resulted in an end to the 'male-breadwinner' form of post war welfare state capitalism in the UK. There was also a perception amongst some participants that components of *some forms* of family life had been undermined, as both parents generally must work to forge a livelihood. The complex sentiments outlined in this chapter particularly around the area's lost industry, insecure work, decline of community spirit, boarded up stores and degeneration of the high street, prevalence of criminality and ASB and sense of a geographical divide demonstrates the breadth of problems in this 'left behind' place, embodying structural obstacles to the Levelling Up agenda that will take sustained economic investment over a lengthy period to ameliorate.

Conclusion

Whilst the Levelling Up strategy acknowledges place-based problems and the need to rebalance a profoundly uneven economy (HM Government, 2022), this chapter documented the scale of structural problems in 'left behind' R&C in Teesside. It demonstrated how these issues are long-running and deeply embedded, tethered to the shift from post-war capitalism to neoliberalism in the late 1970s (Jones, 2019; Martin et al., 2021; Wistow, 2022). As documented, the area's former industrial prowess particularly through ICI and the steelworks continues to shape sentiments especially amongst the older residents, who recalled it fondly in part due to its economic stability, security, and camaraderie. During the neoliberal period deindustrialisation was amplified with Teesside losing ICI in the 1990s and its core steelworks in 2015, impacting detrimentally upon the area particularly in terms of the availability of remunerative jobs. There was a perception amongst the participants that the area possessed an abundance of insecure work. The decline of industrial work and its associated cultures coupled with the growth of market values impacted negatively upon socio-cultural relations in the area, with many respondents highlighting the loss of community spirit and degeneration of the local high streets. Whilst there was a degree of spatial variation, the latter was cast as characterised by empty stores, charity and pound shops and takeaways.

This contextual backdrop was viewed as contributing to the normalisation and prevalence of drug dealing and usage in the local authority area. Such criminality was accompanied by what was perceived as high levels of acquisitive crime such as burglary and shoplifting, as well as ASB particularly by younger people. Whilst remaining attuned to the context under which this occurred, it is undeniable that both crime and ASB were socially corrosive and harmful forces in the area which caused distress to many residents, including an intensifying atmosphere of fear, insecurity and uncertainty particularly in the more deprived neighbourhoods. However, many participants claimed the local police (who are currently under 'special measures') were impotent in addressing these issues. The localised sense of being 'left-behind' was also shaped by a perception that there was a North South divide, though within region and locality spatial variations were also acknowledged. Ultimately, there were various interpretations over what Levelling Up means in practice, though many respondents emphasised the importance of remunerative work and expressed hope that freeports will help create prosperity; but were often cynical of what they perceived to be tokenistic measures like relocating Government buildings to Darlington. The problems presented throughout this chapter are long-standing and embedded; they present structural challenges to Levelling Up that will take a long time to undo.

References

Atkinson, R. (2020). *Alpha City: How London was captured by the super-rich*. Verso.

Ayres, T. (2021). Childhood trauma, problematic drug use and coping. *Deviant Behaviour, 42*(5), 578–599.

Bambra, C., & Garthwaite, K. (2015). Austerity, welfare reform and the English health divide. *Area, 47*(3), 341–343.

BBC. (2021). Teesside faces 'scourge' of drug deaths increase. *BBC*. Retrieved July 6, 2022, from https://www.bbc.co.uk/news/uk-england-tees-58221380

Beatty, C., & Fothergill, S. (2018). Welfare reform in the UK 2010–16: Expectations, outcomes and local impacts. *Social Policy & Administration, 52*(5), 950–968.

Boswell, J., Denham, J., Furlong, J., Killick, A., Ndugga, P., Rek, B., Ryan, M., & Shipp, J. (2022). Place-based politics and nested deprivation in the UK: Beyond cities-towns, 'Two Englands' and the 'Left Behind'. *Journal of Representative Democracy, 58*(2), 169–190.

Cant, C., & Woodcock, J. (2020). Fast Food Shutdown: From disorganisation to action in the service sector. *Capital & Class, 44*(4), 513–521.

Chaytor, S., Gottlieb, G., & Reid, G. (2020). Regional policy and R&D: Evidence, experiments and expectations. *HEPI Report 137*.

Coomber, R., & Moyle, L. (2018). The changing shape of street-level heroin and crack supply in England: Commuting, holidaying and cuckooing drug dealers across 'county lines'. *British Journal of Criminology, 58*(6), 1323–1342.

Ellis, A. (2019). A de-civilizing reversal or system normal? Rising lethal violence in post-recession austerity United Kingdom. *British Journal of Criminology, 59*(4), 862–878.

Etherington, D., Telford, L., Jones, M., Harris, S., & Hubbard, S. (2022). *The pending poverty catastrophe in Stoke-on-Trent: How benefit cuts and the cost-of-living crisis impacts on the poor*. Stoke-on-Trent: Staffordshire University.

Goodwin, M., & Heath, O. (2016). The 2016 referendum, Brexit and the left behind: An aggregate-level analysis of the result. *The Political Quarterly, 87*(3), 323–332.

Gray, M., & Barford, A. (2018). The depths of the cuts: The uneven geography of local government austerity. *Cambridge Journal of Regions, Economy and Society, 11*, 541–563.

Hall, A., Antonopoulos, G., Atkinson, R., & Wyatt, T. (2022). Duty free: Turning the criminological spotlight on special economic zones. *British Journal of Criminology*, 1–18. https://doi.org/10.1093/bjc/azac010

Hall, S., & Winlow, S. (2018). Big trouble or little evils: The ideological struggle over the concept of harm. In A. Boulki & J. Kotze (Eds.), *Zemiology: Reconnecting crime and social harm*. Palgrave Macmillan.

Hall, S., Winlow, S., & Ancrum, C. (2008). *Criminal identities and consumer culture: Crime, exclusion and the new culture of narcissism*. Willan Publishing.

Harvey, D. (2005). *A brief history of neoliberalism*. Oxford University Press.

Harvey, D. (2007). Neoliberalism as creative destruction. *The Annals of the American Academy of Political and Social Science, 610*, 22–47.

HMICFRS. (2019). *Police effectiveness, efficiency and legitimacy 2018/2019: An inspection of Cleveland Police*. HMICFRS.

Hoggart, R. (1957). *The uses of literacy: Aspects of working-class life*. Penguin.

Hudson, R. (2022). 'Levelling up' in post-Brexit United Kingdom: Economic realism or political opportunism? *Local Economy, 0*(0), 1–16.

Hughes, M. (2021). British Steel invests £26m at Skinningrove site. *The Northern Echo*. Retrieved July 4, 2022, from https://www.thenorthernecho.co.uk/news/19535967.british-steel-invests-26m-skinningrove-site/

IPPR. (2022). *State of the North 2021/22: Powering northern excellence*. Manchester: IPPR.

Jones, M. (2019). *Cities and regions in crisis: The political economy of sub national economic development*. Edward Elgar.

Kalantaridis, C. (2010). In-migration, entrepreneurship and rural-urban interdependencies: The case of East Cleveland, North East England. *Journal of Rural Studies, 26*, 418–427.

Kotze, J. (2019). *The myth of the 'crime decline': Exploring change and continuity in crime and harm*. Routledge.

Lake, A. (2018). Neighbourhood food environments: Food choice, foodscapes and planning for health. *Proceedings of the Nutrition Society, 77*, 239–246.

Lloyd, A. (2013). *Labour markets and identity on the post-industrial assembly line.* Routledge.
Lloyd, A. (2018). *The harms of work: An ultra-realist account of the service economy.* Policy Press.
Lloyd, A., Devanney, C., Wattis, L., & Bell, V. (2021). "Just tensions left, right and centre": Assessing the social impact of international migration on deindustrialized locale. *Ethnic and Racial Studies, 44*(15), 2794–2815.
MacDonald, R., Shildrick, T., & Furlong, A. (2020). 'Cycles of disadvantage' revisited: Young people, families and poverty across generations. *Journal of Youth Studies, 23*(1), 12–27.
MacKinnon, D., Kempton, L., O'Brien, P., Ormerod, E., Pike, A., & Tomaney, J. (2022). Reframing urban and regional 'development' for 'left behind' places. *Cambridge Journal of Regions, Economy and Society, 15*, 39–56.
MacLeod, G., & Jones, M. (2018). Explaining 'Brexit capital': Uneven development and the austerity state. *Space and Polity, 22*(2), 111–136.
Martin, R. (1988). The political economy of Britain's North-South divide. *The Royal Geographical Society, 13*(4), 389–418.
Martin, R. (2021). Rebuilding the economy from the Covid crisis: Time to rethink regional studies? *Regional Studies, Regional Science, 8*(1), 143–161.
Martin, R., Gardiner, B., Pike, A., Sunley, P., & Tyler, P. (2021). *Levelling Up left behind places.* Routledge.
Measham, F., Newcombe, R., & Parker, H. (1994). The normalization of recreational drug use amongst young people in North-West England. *British Journal of Sociology, 45*(2), 287–312.
Office for National Statistics [ONS]. (2021a). Labour Market Profile—Redcar & Cleveland. ONS.
Office for National Statistics [ONS]. (2021b). Crime in England and Wales: Police Force Area data tables. ONS.
Osborne, H. (2021). Average London house price exceeds £500,000 for first time. *The Guardian*. Retrieved July 20, 2022, from https://www.theguardian.com/money/2021/jan/20/average-london-house-price-exceeds-500000-for-first-time-covid
Pickett, K., & Wilkinson, R. (2010). *The spirit level: Why equality is better for everyone.* Bloomsbury Press.
Raymen, T. (2016). Designing-in crime by designing-out the social? Situational crime prevention and the intensification of harmful subjectivities. *British Journal of Criminology, 56*, 497–514.
Statista. (2019). Share of charity shops in high streets in the United Kingdom (UK) in 2019, by region*. Statista.
Streeck, W. G. (2016). *How will capitalism end?* Verso.

Tees Valley Mayor. (n.d.). Chemicals and process. Tees Valley Combined Authority. Retrieved July 4, 2022, from https://teesvalley-ca.gov.uk/business/key-sectors/chemicals-and-process/

Telford, L. (2022a). *English nationalism and its ghost towns*. Routledge.

Telford, L. (2022b). "There is nothing there": Deindustrialization and loss in a coastal town. *Competition & Change, 26*(2), 197–214.

Telford, L., Bushell, M., & Hodgkinson, O. (2022). Passport to neoliberal normality? A critical exploration of COVID-19 vaccine passports. *Journal of Contemporary Crime, Harm, Ethics, 2*(1), 42–61.

Telford, L., & Wistow, J. (2020). Brexit and the working class on Teesside: Moving beyond reductionism. *Capital & Class, 44*(4), 553–572.

Tomaney, J., & Pike, A. (2020). Levelling Up? *The Political Quarterly, 91*(1), 43–48.

Treadwell, J., Ancrum, C., & Kelly, C. (2020). Taxing times: Inter-criminal victimization and drug robbery amongst the English professional criminal milieu. *Deviant Behavior, 41*(1), 57–69.

Wakeman, S. (2021). Doing autoethnographic drugs research: Some notes from the field. *International Journal of Drug Policy*. Online First: https://doi.org/10.1016/j.drugpo.2021.103504

Walkerdine, V. (2020). 'No-one listens to us': Post-truth, affect and Brexit. *Qualitative Research in Psychology, 17*(1), 143–158.

Warren, J. (2018). *Industrial Teesside: Lives and legacies*. Palgrave Macmillan.

Williamson, M. (2012). *Life at the yard: Memories of working at Smith's Dock South Bank*. Teesside Industrial Memories Project.

Winlow, S., & Hall, S. (2006). *Violent night: Urban leisure and contemporary culture*. Berg Publishers.

Winlow, S., & Hall, S. (2013). *Rethinking social exclusion: The end of the social?* SAGE.

Wistow, J. (2022). *Social policy, political economy, and the social contract*. Policy Press.

CHAPTER 6

Conclusion: Scanning the Future

Abstract The book's concluding chapter briefly summarises the book's key contributions, particularly its qualitative account of Levelling Up a 'left behind' place. It then speculates on the future, documenting how the current crises provide fertile grounds in the UK for shifting into a post-neoliberal world and achieve significant Levelling Up. It explores how it is possible that a Green Industrial Revolution and deglobalization will be core components of the UK political economy in the upcoming years, as well as how a state funded job guarantee could potentially be transformative for 'left behind' places.

Keywords Post-neoliberal • Deglobalization • Levelling Up

Our small-scale qualitative research study sought to contribute to debates on 'left behind' places and the Levelling Up agenda, particularly through qualitative data primarily gathered from one of the UK's most 'left behind' local authorities (Martin et al., 2021). At first, the book summarised the Levelling Up strategy and the methodological underpinning of our research approach. This included 30 semi-structured interviews involving five Directors of Regeneration working in 'left behind' places, as well as with 25 residents of 'left behind' R&C in Teesside. The introduction also outlined the complex systems frame of reference that was utilised as an overarching framework, shaping our data collection and conceptual view

L. Telford, J. Wistow, *Levelling Up the UK Economy*,
https://doi.org/10.1007/978-3-031-17507-7_6

of the Levelling Up 'left behind' places problem. This framework was occasionally deployed to enhance the analysis. Such an approach is explicitly tied to an understanding of systems that includes the role of the political economy in shaping the meso and micro levels, leading to the spatial disparities that the government is seeking to address. If the political economy has a causal relationship to spatial inequalities, it follows that it is also a tool or lever for addressing these. As we saw in the Introduction Chapter, the then Prime Minister—Boris Johnson—was clear when launching the Levelling Up agenda that he was setting a high bar for the equalisation of opportunities and outcomes relative to place of birth. However, it is far less clear that the ambition associated with the agenda has stretched to intervening sufficiently at the macro and meso levels of complex systems in which these levers are located.

Chapter 2 provided a historical context of the different phases of capitalism from the mid nineteenth century to the present. It documented these changes at different levels—macro, meso and micro—and the implications for place. In particular, a historical overview of R&C and the broader Teesside subregion was provided as context for the case study in Chap. 5. Crises in the first half of the twentieth century such as the First World War, 1929 Great Depression, mass unemployment, chronic economic insecurity, and the march of fascism and communism led to an anomalist phase in capitalism in which successive governments played a central role in intervening in markets to mitigate their tendency to generate inequalities (Mitchell & Fazi, 2017; Streeck, 2016). Successive governments maintained a commitment to what became known as the 'class compromise', with full employment a key political objective (Judt, 2010; Telford, 2022). Relative economic stability and social security were stitched into the fabric of working-class life. However, crises in the 1970s including capital's declining profitability, a global oil crisis, high levels of inflation and economic stagnation enabled the system to change again (Mitchell & Fazi, 2017). The *phase shift* from a Keynesian and Fordist political economy to neoliberalism in the late 1970s engendered a change in type of the system (Wistow, 2022), with the deindustrialisation of the UK economy and abandonment of full employment as a policy commitment forming key parts of this restructuring (MacLeavy, 2019; Wistow, 2022).

Whilst London and the Southeast were able to reinvent themselves partially in light of the Big Bang deregulation of the British economy in the 1980s (Jones, 2019; Wistow, 2022), many other areas particularly in the

North gradually fell behind. This led to a profoundly spatially uneven society, which has been further exposed in light of more recent events, such as the 2008 global financial crash and the imposition of austerity. This was felt most harshly in relatively deprived and post-industrial localities across the North, aggravating place-based inequalities and forming a contributory factor to the rise of political discontent vis a vis Brexit and the 2019 collapse of the Red Wall (MacLeavy & Jones, 2021). These political earthquakes resulted in increased debate about so-called 'left behind' places, which have failed to keep pace with other locales and regions under neoliberalism (Martin, 2021).

Chapter 3 explored how the Levelling Up strategy proposes to remedy the UK's regional inequalities particularly the problems of 'left behind' locales. It outlined discussions on Levelling Up prior to the LUWP including how it was described as ideologically ambiguous, conceptually nebulous and lacking clear goals and measures. It then explored the LUWP and how it was framed within a complex systems frame of reference, as well as its contradictions considering the Conservative Party's ideological position; one that favours market forces and possessive individualism over redistributive economic and social policy. It also deployed findings from interviews with the Directors of Regeneration from the 'Northern Core City'; a 'Northern Other City'; a 'Northern Large Town'; a 'Midlands Medium Town'; and a 'London Borough'.

This laid the foundations for Chap. 4, which focused on a specific 'entry point'—local government—to the Levelling Up agenda and the complex systems cutting across this. Findings from interviews with Directors of Regeneration were further used to frame issues identified in the wider literature about local democracy and devolution, local government in local governance systems and the role of local government in relation to the Levelling Up agenda. In particular, local government has faced over 40 years of relative decline as its functional responsibilities have been contracted out and its funding, especially during the austerity years, was dramatically reduced. This is a far from promising context to deliver place-based policy. Given the weaknesses of the Levelling Up agenda as a place-based policy, as discussed in Chap. 3, challenges in addressing spatial disparities are further compounded.

Chapter 5 offered an empirical case study of the challenges to Levelling Up R&C in Teesside. It documented how the socio-economic decline of this 'left behind' locale is long standing and embedded, bound to the shift from post-war capitalism to neoliberalism (Telford, 2022). As the local economy deindustrialised, more insecure, uncertain and exploitative forms of work proliferated (Lloyd, 2018; Telford & Wistow, 2020; Warren, 2018). Such economic restructuring meant the area had declined for several decades, involving the diminishment of a sense of community and proliferation of boarded up stores. Participants emphasised how their high streets were populated by pound shops, charity shops and fast-food outlets. This was accompanied by criminality and ASB, which were core localised problems and sources of social distress.

The residents possessed various interpretations over what Levelling Up means in practice and how it can address the area's problems. This included the importance of remunerative work, while others expressed cautious optimism over if the place's free port would be a success in delivering jobs and local economic growth. However, evidence indicates that rather than providing well-paid jobs and local prosperity, free ports are sites of wealth extraction and harm (see: Hall et al., 2022). Other respondents were more cynical, suggesting the task of Levelling Up the local authority area to more affluent places is too difficult given the extent to which it has been 'left behind' over a long period of time. Notwithstanding, all participants acknowledged the scale and nature of the problems for Levelling Up, embodying policy challenges that will take a long time to ameliorate. Our book now closes with a look to the future, largely exploring the importance of shifting to a post-neoliberal political economy.

Moving into a Post-Neoliberal World?

This book has spent some time primarily documenting where we have come from –considering and analysing the Levelling Up strategy, findings from interviews with Directors of Regeneration around the agenda and local governance, as well as the problems in a 'left behind' place—meaning it is fitting that we conclude by scanning some possible future developments. It is clear from our research—particularly the interviews with Directors of Regeneration and working-class residents—that the problems in 'left behind' communities are deep rooted and ingrained. It is also evident that under the government's current Levelling Up funding proposals, the resources required to achieve the LUWP's ambitions have not been allocated (Hudson, 2022; MacKinnon et al., 2022). As mentioned, Martin et al. (2021) argue that it would require resources akin to Germany's Aufbau Ost programme in light of German reunification, which the current funding commitments of the Levelling Up agenda pale in comparison to. This is essential as Martin et al. (2021 p.108) highlight:

> "The scale and nature of the UK's contemporary left behind places problem are such that only a transformative shift in policy model and resource commitment of historic proportions are likely to achieve the levelling up ambition that has become such a prominent theme in the current government's political declarations".

Furthermore, changes to the system logic at the macro and meso levels may help to create a more tractable context for policies designed to tackle spatial disparities. For example, Wistow (2022) argued that measures taken to make the political economy more redistributive (such as having a high-level policy goal to reduce social inequalities to the level of the 1970s, through a range of land, wealth, income and corporation taxes) and returning utilities and services to public ownership and state delivery where possible, are necessary for a more equitable social contract. He (2022 p.151) argued that changes of this kind and scale are required to recalibrate complex systems, 'in order to make them more tractable to well-meaning policy objectives'. The Levelling Up agenda is the type of policy initiative that should be seeking to contribute to a more equitable society; but it is 'tagged onto' rather than fundamentally challenging the make-up of the political economy.

Given the myriad cost-of-living crises, transformative change takes on increased urgency and importance. The accumulation of crises has historically served to provide opportunities for restructuring economies and societies. For example, Martin (2021) and Telford (2022) highlighted that the 2008 global financial crisis, the rise of political dissatisfaction and COVID-19 pandemic provide fertile grounds for historic change, which may be heightened further in light of the war in Ukraine and cost-of-living crisis. These forces have diminished the legitimacy of a neoliberal and rentier political economy to deliver positive economic and social outcomes, particularly for those living in 'left behind' locales. Indeed, many commentators have highlighted the potential for a shift into a post-neoliberal world involving a Green Industrial Revolution/New Deal or a Fourth Industrial Revolution (Briggs et al. 2021; Etherington et al., 2022; Martin, 2021; Martin et al., 2021; Russell et al., 2022), meaning Levelling Up should be more aligned with other emerging agendas (Martin, 2021). For example, R&C could see somewhat of a renaissance through being a key part of the energy transition to tackle global warming vis a vis a Green Industrial Revolution, given plans for Teesside to be the UK's first decarbonised industrial cluster (Telford, 2022).

This is particularly urgent with inflation predicted to reach 18% in early 2023, which would be the highest level in nearly fifty years (Lawson & Mason, 2022). The weaknesses of neoliberal globalisation through, for example, deindustrialisation, hollowed out state capacity, and the offshoring of much energy manufacturing, makes the UK economy far less resilient to external shocks and has been exposed, yet again. Western sanctions in response to Russia waging war in Ukraine has meant Russia responded by reducing their gas flow to Europe via the Nord Stream 1 pipeline, massively increasing prices for the consumer as demand surges (Fazi, 2022). After decades of wage stagnation and rising private indebtedness under neoliberalism, millions of households do not have the disposable income available to absorb this financial pressure. 'Left behind' localities, in particular, have higher levels of low-income households, making them more vulnerable to price rises. In effect, without significant state support place-based inequalities will amplify as millions of people will fall into poverty (Brewer et al., 2022), further falling behind rather than Levelling Up. The scale and depth of the crisis means bold change and policy experimentation is required, which has also been championed by scholars regarding Levelling Up 'left behind' places (Etherington et al., 2022; MacKinnon et al., 2022; Martin, 2021; Martin et al., 2021).

Whilst it is debatable what the implications of the recent appointment of Rishi Sunak as the Conservative Prime Minister means for Levelling Up, it is essential austerity is not reintroduced due to its depressive economic impact particularly in already struggling 'left behind' localities (Martin, 2021). It is conceivable that the language of fiscal sensibility may diminish as it becomes clear that only significant state intervention in markets and investment can steer the UK through the turbulent times that lie ahead and achieve significant Levelling Up. One such means is through rebuilding the foundational economy that neoliberalism engaged in 'wrecking' (Russell et al., 2022, p. 1072). This includes nationalising the social infrastructure of everyday life such as care, education, health, food production and energy. Such rebuilding would ensure that these core utilities and services are operated for the *common good* rather than for private gain. There are strong parallels here with MacKinnon et al.'s (2022) emphasis on how redeveloping 'left behind' locales should focus upon five principles of human need including the need to belong and feel secure, the need to feel in control of one's life and the need to support one's family.

Indeed, state spending during the COVID-19 pandemic demonstrated that money is available for societal restructuring. To use a well-worn analogy, the magic money tree was shaken, and vast sums of public money were spent upon supporting businesses, individuals, families and communities particularly through the furlough schemes to prevent mass unemployment and total economic disintegration (Briggs et al. 2021; Gerbaudo, 2021). A deeply ingrained neoliberal shibboleth—that the state is fiscally constrained and like a household in its spending—was somewhat lacerated, with the usual question of 'how to pay for it' not being considered (Briggs et al. 2021). The idea that the state is like a household chimes with many people, since we know that if we borrow money, we must repay the debt otherwise we risk financial consequences such as an inability to pay bills, higher interest rates when borrowing and insolvency. Proponents of Modern Monetary Theory (MMT), however, suggest that the state and household dictum is the foundational economic myth of neoliberalism since sovereign states like the UK are the issuers of currency and not the users (Kelton, 2020; Mitchell & Fazi, 2017; Tcherneva, 2020). The Bank of England creates money electronically, meaning there is no need to 'balance the budget' and reduce the deficit. In fact, deficits are important to a healthy economy, stimulating economic growth and the resources required such as the labour supply to utilise the state's investment (see: Kelton, 2020; Mitchell & Fazi, 2017; Tcherneva, 2020).

MMT's founding thinkers tend to be from the political Left and one of their closely aligned social policies is a state funded job guarantee (Kelton, 2020; Mitchell & Fazi, 2017; Tcherneva, 2020). Given how the loss of relatively remunerative industrial jobs formed the background to an array of cultural problems like crime and ASB, as well as the higher than national average levels of joblessness in 'left behind' places (MacKinnon et al., 2022; Martin et al., 2021; Sandbu 2020; Tomaney et al., 2021), it is difficult to underestimate the importance of well-paid work to resurrect 'left behind' areas. Our data alluded to the indignities of insecure work, as well as the difficulties many people face in finding well-paid employment in 'left behind' locales (also see: Etherington et al., 2022; MacKinnon et al., 2022; Martin et al., 2021; Sandbu 2020; Tomaney et al., 2021). A job guarantee would alter the relation between the state and citizens, shifting it away from governing for capital and instead protecting the socio-economic interests of ordinary people. It could be set at a living wage, driving up wages in the private sector and providing an alternative to low-pay and insecure jobs. Therefore, it would help to bolster the minimum standard of income and working conditions, while dispelling the neoliberal notion that joblessness and communal disintegration are inevitable (see: Kelton, 2020; Mitchell & Fazi, 2017; Tcherneva, 2020). Given the significance placed on waged work to 'left behind' places like R&C there may be wider benefits around pride in place and the amelioration of some of the social problems concentrated here, including high levels of socially corrosive crime and ASB.

Gerbaudo (2021) highlighted the potential for states to provide a defence against external threats amid potential moves towards 'deglobalization'. Given the significance of the price of energy in the current cost of living crisis, energy independence may be a forerunner in this debate. Gerbaudo (2021) argues that it would provide more economic resilience in an increasingly turbulent world, enabling states to award more protection for their citizens and control over the economy. Government officials have spoken about the need for energy sovereignty in the years ahead particularly through the increased production of nuclear power, solar energy and onshore wind development (Stewart, 2022), potentially creating relatively remunerative and long-term jobs of the future. Of course, there is no guarantee that this transition will be spatially just and equitable, since capitalism creates uneven spatial development by its very nature and benefits certain places at the expense of others (Hudson, 2022; Jones, 2019; Streeck, 2016). Therefore, it is important the green economy of the future 'does not just favour existing technologically innovative places or the richer sections of society, but provides a means of reviving and rebuilding places whose carbon-based economies will need to be reorientated the most' (Martin, 2021 p.150). Indeed, until resources and funds of historical magnitude are allocated to achieve the LUWP's aspirations, aiding a political economic shift away from neoliberalism and its language of fiscal constraint, 'left behind' communities like R&C and their various problems will persist. Once we emerge out of the current crises, we cannot afford to continue with a failed economic model; the UK's political class must be committed to a phase shift in the political economy. Otherwise, the UK economy will in all likelihood continue to leave places behind rather than Level them Up.

References

Brewer, M., Fry, E., Handscomb, K., & Marshall, J. (2022). *A chilling crisis*. Resolution Foundation.

Etherington, D., Jones, M., & Telford, L. (2022). Challenges to Levelling Up: Post-COVID precarity in "left behind" Stoke-on-Trent. *Frontiers in Political Science*. Online First: https://www.frontiersin.org/articles/10.3389/fpos.2022.1033525/full.

Fazi, T. (2022). Europe has lost the energy war. UnHerd. Retrieved 24 August, 2022 https://unherd.com/2022/08/europe-has-lost-the-energy-war/

Gerbaudo, P. (2021). *The great recoil*. Verso.

Hall, A., Antonopoulos, G., Atkinson, R., & Wyatt, T. (2022). Duty free: Turning the criminological spotlight on special economic zones. *British Journal of Criminology, 1*, 1–18.

Hudson, R. (2022). 'Levelling up' in post-Brexit United Kingdom: Economic realism or political opportunism? *Local Economy, 37*(1–2), 50–65.

Judt, T. (2010). *Post war: A history of Europe since 1945*. Penguin.

Jones, M. (2019). *Cities and regions in crisis: The political economy of sub National Economic Development*. Edward Elgar.

Kelton, S. (2020). *The deficit myth*. Public Affairs.

Lawson, A., & Mason, R. (2022) UK inflation will hit 18% in early 2023, says leading bank Citi. The Guardian. Retrieved 30 August, 2022, from https://www.theguardian.com/business/2022/aug/22/uk-inflation-will-hit-18-per-cent-in-early-2023-says-leading-bank-citi-gas-electricity

Lloyd, A. (2018). *The harms of work*. Policy Press.

MacKinnon, D., Kempton, L., O'Brien, P., Ormerod, E., Pike, A., & Tomaney, J. (2022). Reframing urban and regional 'development' for 'left behind' places. *Cambridge Journal of Regions, Economy and Society, 15*, 39–56.

MacLeavy, J. (2019). Neoliberalism and the new political crisis in the west. *Ephemera, 19*(3), 627–640.

MacLeavy, J., & Jones, M. (2021). Brexit as Britain in decline and its crises (revisited). *The Political Quarterly, 92*(3), 444–452.

Martin, R. (2021). Rebuilding the economy from the Covid crisis: Time to rethink regional studies? *Regional Studies, Regional Science, 8*(1), 143–161.

Martin, R., Gardiner, B., Pike, A., Sunley, P., & Tyler, P. (2021). *Levelling up left behind places: The scale and nature of the economic and policy challenge*. Routledge.

Mitchell, W., & Fazi, T. (2017). *Reclaiming the state*. Pluto Press.

Russell, B., Beel, D., Jones, I., & Jones, M. (2022). Placing the foundational economy: An emerging discourse for post-neoliberal economic development. *EPA: Economy and Space, 54*(6), 1069–1085.

Stewart, H. (2022). UK energy independence strategy: what are the cabinet divisions. The Guardian. Retrieved 31 August, 2022, from https://www.theguardian.com/environment/2022/mar/28/cabinet-divisions-over-uk-energy-independence-strategy-boris-johnson

Streeck, W. G. (2016). *How will capitalism end?* Verso.

Tcherneva, P. (2020). *The case for a job guarantee*. Polity Press.
Telford, L., & Wistow, J. (2020). Brexit and the working class on Teesside: Moving beyond reductionism. *Capital & Class, 44*(4), 553–572.
Telford, L. (2022). *English nationalism and its ghost towns*. Routledge.
Tomaney, J., Natarajan, L., & Sutcliffe-Braithwaite, F. (2021). *Sacriston: Towards a deeper understanding of place*. University College London.
Warren, J. (2018). *Industrial Teesside: Lives and legacies*. Palgrave Macmillan.
Wistow, J. (2022). *Social policy, political economy and the social contract*. Policy Press.

Ingram Content Group UK Ltd.
Milton Keynes UK
UKHW010739180523
421954UK00001B/29